TOPICS IN COMPLEX FUNCTION THEORY

INTERSCIENCE TRACTS
IN PURE AND APPLIED MATHEMATICS

Editors: L. BERS · R. COURANT · J. J. STOKER

1. *D. Montgomery and L. Zippin*—Topological Transformation Groups
2. *Fritz John*—Plane Waves and Spherical Means Applied to Partial Differential Equations
3. *E. Artin*—Geometric Algebra
4. *R. D. Richtmyer*—Difference Methods for Initial-Value Problems
5. *Serge Lang*—Introduction to Algebraic Geometry
6. *Herbert Busemann*—Convex Surfaces
7. *Serge Lang*—Abelian Varieties
8. *S. M. Ulam*—A Collection of Mathematical Problems
9. *I. M. Gel'fand*—Lectures of Linear Algebra
10. *Nathan Jacobson*—Lie Algebras
11. *Serge Lang*—Diophantine Geometry
12. *Walter Rudin*—Fourier Analysis on Groups
13. *Masayoshi Nagata*—Local Rings
14. *Ivan Niven*—Diophantine Approximations
15. *S. Kobayashi and K. Nomizu*—Foundations of Differential Geometry. In two volumes
16. *J. Plemelj*—Problems in the Sense of Riemann and Klein
17. *Richard Cohn*—Difference Algebra
18. *Henry B. Mann*—Addition Theorems: The Addition Theorems of Group Theory and Number Theory
19. *Robert Ash*—Information Theory
20. *W. Magnus and S. Winkler*—Hill's Equation
21. *Zellig Harris*—Mathematical Operations in Linguistics
22. *C. Corduneanu*—Almost-Periodic Functions
23. *Richard S. Bucy and Peter D. Joseph*—Filtering and Its Applications
24. *Paulo Ribenboim*—Rings and Modules (in preparation)
25. *C. L. Siegel*—Topics in Complex Function Theory—Vol. I Elliptic Functions and Uniformization Theory

TOPICS
IN COMPLEX
FUNCTION THEORY

BY

C. L. SIEGEL

VOL. I

Elliptic Functions and Uniformization Theory

TRANSLATED FROM THE ORIGINAL GERMAN BY
A. SHENITZER
Adelphi University, Garden City

AND

D. SOLITAR
York University, Toronto

WILEY—INTERSCIENCE

A DIVISION OF JOHN WILEY & SONS

NEW YORK · LONDON · SYDNEY · TORONTO

Copyright © 1969 by John Wiley & Sons, Inc.

All rights reserved. No part of this book may
be reproduced by any means, nor transmitted,
nor translated into a machine language without
the written permission of the publisher.

10   9   8   7   6   5   4   3   2   1

Library of Congress Catalog Card Number: 69-19931
SBN 471 79070 2 ✓

Printed in the United States of America

The present version of my lectures in function theory, which I delivered at the university of Göttingen beginning in the Fall of 1964, is based on notes prepared from my lectures some ten years ago by E. Gottschling. I supplemented and improved these notes to some extent but, as opposed to other authors, I decided against the adoption of the terminology and abstract formulations that have since become fashionable.

C. L. Siegel

Preface to the English edition

Topics in Complex Function Theory consists of three parts corresponding to three consecutive courses taught by Professor Carl L. Siegel at the University of Göttingen. Lecture notes, in German, were taken by Erhard Gottschling and (for the second half of Part III) by Helmut Klingen. They were made available in mimeographed form by the Mathematical Institute of the University of Göttingen and were later revised by Professor Siegel, who then gave me permission to arrange for the publication of an English translation. The text was solicited for "Interscience Tracts" by Professor Lipman Bers. That Parts I and II can now appear is due to the dedicated efforts of Professor Abe Shenitzer who produced a complete first draft in English. He was assisted in questions of terminology and style and in checking the correctness of the translation by his colleagues Donald Solitar and Trueman MacHenry at Adelphi University and by Mr. Marvin Tretkoff at New York University. In some cases the translators have added notes concerning the use of certain terms, but these notes are marked as such.

A translation of Part III on "Abelian Functions and Modular Functions in Several Variables" is in preparation. After its completion an English text written by a great mathematician of our time will present one of the most fascinating fields of mathematics.

New York   *Wilhelm Magnus*
March 1969   *New York University*

Contents

CHAPTER 1

Elliptic functions

1. Doubling the arc of a lemniscate 1
2. The Euler addition theorem 7
3. Analytic continuation 11
4. Riemann regions 16
5. The Riemann surface of the function

$$\sqrt{a_0 z^4 + a_1 z^3 + a_2 z^2 + a_3 z + a_4}$$ . . . 22

6. The elliptic integral of the first kind 29
7. The inverse function 31
8. The covering surface 37
9. The periods 40
10. The period parallelogram 48
11. The $\wp$-function 56
12. Partial fractions expansion of the $\wp$-functions 61
13. The inversion problem 66
14. The field of elliptic functions 70
15. The addition theorem 80
16. Degenerate elliptic functions 85

CHAPTER 2

Uniformization

1. Algebraic functions 90
2. Compact Riemann regions 104
3. The fundamental group 113
4. Invariance of the genus 124
5. The Poisson integral 129
6. The Dirichlet integral 138
7. Preliminaries for the mapping theorem 147
8. Construction of a harmonic function with minimal property . 154
9. The mapping theorem 168
10. Uniformization of algebraic functions 178

Index 185

TOPICS IN COMPLEX FUNCTION THEORY

1

Elliptic Functions

1. Doubling the arc of a lemniscate

The theory of elliptic functions developed in the 19th century has a long history. It begins with a discovery, which we are about to discuss, of a remarkable property of the arc of a lemniscate made in 1718 by the Italian count Fagnano.

As is well known, a lemniscate is the locus of a point η in a plane such that the product of its distances from two fixed points has constant value c^2. Let $2a$ be the fixed distance between the two fixed points η_1 and η_2. We choose a rectangular coordinate system in which the points η_1 and η_2 have coordinates $(-a, 0)$ and $(a, 0)$. If r_1, r_2, and r are the distances from a point η on the lemniscate with coordinates (x, y) to η_1, η_2, and the origin (Figure 1), then

(1)
$$r^2 = x^2 + y^2$$

and

$$r_1^2 = (x + a)^2 + y^2 = r^2 + a^2 + 2ax,$$
$$r_2^2 = (x - a)^2 + y^2 = r^2 + a^2 - 2ax.$$

If we multiply the last two equations and bear in mind that $r_1 r_2 = c^2$, then we obtain

(2)
$$r^4 + 2a^2 r^2 + a^4 - 4a^2 x^2 = c^4.$$

By varying the value of c we obtain a class of lemniscates. In this class we consider the lemniscates which pass through the origin and have the form of a horizontal figure 8. This means that $c = a$. It is convenient to put $2a^2 = 1$. Then the distance between the points η_1 and η_2 is $\sqrt{2}$ and the equation (2) becomes

(3)
$$2x^2 = r^2 + r^4.$$

Using (1) to eliminate x from (3) we obtain

(4)
$$2y^2 = r^2 - r^4.$$

After extraction of roots, (3) and (4) yield a parametric representation of the lemniscate with radius vector r as parameter.

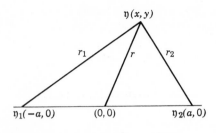

Figure 1

We shall now compute the length of an arc of the lemniscate. Since the curve is symmetric with respect to both coordinate axes, we can restrict ourselves to an arc from the origin to a point in the first quadrant (Figure 2). In order to compute the corresponding arc length s, we regard r as an independent variable which varies over the interval $0 \le r \le 1$. (We use a dot to denote differentiation with respect to r.) In view of (3) and (4) we have

$$2x\dot{x} = r + 2r^3,$$
$$2y\dot{y} = r - 2r^3,$$
$$(2xy)^2\dot{s}^2 = (2xy)^2(\dot{x}^2 + \dot{y}^2) = y^2(r + 2r^3)^2 + x^2(r - 2r^3)^2$$
$$= \frac{r^2 - r^4}{2}(r^2 + 4r^4 + 4r^6) + \frac{r^2 + r^4}{2}(r^2 - 4r^4 + 4r^6) = r^4;$$

further,

$$(2xy)^2 = (r^2 + r^4)(r^2 - r^4) = r^4(1 - r^4),$$

so that

$$(1 - r^4)\dot{s}^2 = 1, \qquad \frac{ds}{dr} = \frac{1}{\sqrt{1 - r^4}}.$$

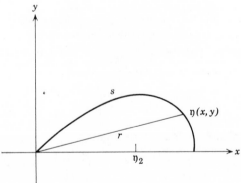

Figure 2

If, beginning at the origin, we trace the arc of the lemniscate through the entire first quadrant, then r takes on every value in the internal (0, 1) exactly once. Integration yields the relation

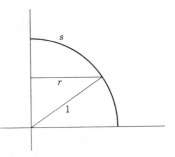

(5)
$$s = s(r) = \int_0^r \frac{dr}{\sqrt{1 - r^4}} \qquad (0 \le r \le 1),$$

where we take the positive square root in the integrand.

Figure 3

In view of the relation just established, the integral $s = s(r)$ in (5) is referred to as the *lemniscatic integral*. The integrand contains a square root of a polynomial of fourth degree. It appears that Fagnano attempted to rationalize this integrand by means of a suitable substitution, and here he may have had in mind the following situation: If, in place of (5), we consider the analogous formula for the arc of a circle, namely (Figure 3),

(6)
$$\arcsin r = \int_0^r \frac{dr}{\sqrt{1 - r^2}} \qquad (0 \le r \le 1),$$

then, as is well known, it is possible to rationalize the integrand in the latter integral by means of the substitution

(7)
$$r = \frac{2t}{1 + t^2} \qquad (0 \le t \le 1);$$

in fact,

$$\sqrt{1 - r^2} = \frac{1 - t^2}{1 + t^2}, \qquad \frac{dr}{dt} = 2\frac{1 - t^2}{(1 + t^2)^2}, \qquad \frac{dr}{\sqrt{1 - r^2}} = \frac{2dt}{1 + t}.$$

Now in the lemniscatic integral we have $\sqrt{1 - r^4}$ instead of $\sqrt{1 - r^2}$. This suggests the use of the analogous substitution

(8)
$$r^2 = \frac{2t^2}{1 + t^4},$$

which yields

$$\sqrt{1 - r^4} = \frac{1 - t^4}{1 + t^4}, \qquad r\sqrt{1 - r^4} = \sqrt{2}\,\frac{t(1 - t^4)}{(1 + t^4)^{3/2}},$$

$$r\frac{dr}{dt} = \frac{2t(1 - t^4)}{(1 + t^4)^2}, \qquad \frac{dr}{\sqrt{1 - r^4}} = \sqrt{2}\,\frac{dt}{\sqrt{1 + t^4}}.$$

The relation (8) effects a monotonic mapping of the interval $0 \le t \le 1$ onto the interval $0 \le r \le 1$. It follows that

$$(9) \qquad \int_0^r \frac{dr}{\sqrt{1 - r^4}} = \sqrt{2} \int_0^t \frac{dt}{\sqrt{1 + t^4}} \qquad (0 \le r \le 1);$$

here t is the solution of the equation (8) which lies in the interval $0 \le t \le 1$.

The substitution (8) has not rationalized the integrand of the lemniscatic integral and seems to offer no advantages whatsoever. This is not so, for, as Fagnano recognized, the substitution leads to a geometric construction for the doubling of a given arc of the lemniscate. Namely, if we treat the integrand on the right-hand side of (9) in an analogous manner by using the substitution

$$(10) \qquad t^2 = \frac{2u^2}{1 - u^4},$$

then

$$\sqrt{1 + t^4} = \frac{1 + u^4}{1 - u^4}, \qquad t\sqrt{1 + t^4} = \sqrt{2}\,\frac{u(1 + u^4)}{(1 - u^4)^{3/2}},$$

$$t\,\frac{dt}{du} = \frac{2u(1 + u^4)}{(1 - u^4)^2}, \qquad \frac{dt}{\sqrt{1 + t^4}} = \sqrt{2}\,\frac{du}{\sqrt{1 - u^4}}.$$

As u varies from 0 to 1, t varies monotonically from 0 to ∞ with $t = 0$ for $u = 0$. Hence

$$(11) \qquad \int_0^t \frac{dt}{\sqrt{1 + t^4}} = \sqrt{2} \int_0^u \frac{du}{\sqrt{1 - u^4}}.$$

In view of (9) we now have

$$(12) \qquad \int_0^r \frac{dr}{\sqrt{1 - r^4}} = 2 \int_0^u \frac{du}{\sqrt{1 - u^4}}.$$

If we think of u as a radius vector, then the integral

$$(13) \qquad \sigma = s(u) = \int_0^u \frac{du}{\sqrt{1 - u^4}}$$

again represents the length of the arc of a lemniscate from the origin to a point (ξ, η) in the first quadrant whose coordinates are determined from the equations

$$(14) \qquad 2\xi^2 = u^2 + u^4, \qquad 2\eta^2 = u^2 - u^4.$$

In view of (5), (12), and (13) we have

$$(15) \qquad s = s(r) = 2s(u) = 2\sigma.$$

What matters is that there is a simple algebraic connection between the upper limits r and u of the lemniscatic integrals. Namely, in view of (8) and (10),

$$(16) \qquad r^2 = \frac{4u^2(1 - u^4)}{(1 + u^4)^2},$$

so that r^2 is a rational function of $u^2 = \xi^2 + \eta^2$. Thus, given a point (ξ, η), we can determine the coordinates x and y from (3) and (4) by computing two square roots of rational functions of $\xi^2 + \eta^2$, and this can be done by means of elementary geometric constructions, that is, using ruler and compass alone. In view of (15), the length s of the lemniscatic arc with end point (x, y) is double the length σ of the lemniscatic arc with end point (ξ, η).

Conversely, it is possible to go from (x, y) to (ξ, η) by means of elementary geometric constructions. In fact, in view of (1), r^2 is a rational function of x and y; t^2 is obtained from (8) by solving the quadratic equation

$$(t^2)^2 - \frac{2}{r^2} t^2 + 1 = 0,$$

namely,

$$(17) \qquad t^2 = r^{-2} - \sqrt{r^{-4} - 1} = \frac{1}{r^{-2} + \sqrt{r^{-4} - 1}} \leq 1;$$

u^2 is obtained from (10) by solving the quadratic equation

$$(u^2)^2 + \frac{2}{t^2} u^2 - 1 = 0,$$

namely,

$$(18) \qquad u^2 = \sqrt{t^{-4} + 1} - t^{-2} = \frac{1}{t^{-2} + \sqrt{t^{-4} + 1}} < 1;$$

finally, ξ, η are obtained from (14) by an additional extraction of square roots. All in all, we have shown how to double and halve a lemniscatic arc given by its end point using ruler and compass alone. This was the discovery of Fagnano (Figure 4).

Observe that in the above considerations we restricted ourselves to lemniscatic arcs in the first quadrant. This means that the arc to be doubled must be chosen so small that the end point of the doubled arc still lies in the first quadrant. Since the intervals for r, t, u are connected by monotonic relations, it follows that the largest admissible value of u is obtained from (17) and (18) by taking $r = 1$; then $t^2 = 1$, $u^2 = \sqrt{2} - 1$. In view of (14) we obtain for the corresponding coordinates

$$2\xi^2 = u^2 + u^4 = u^2(1 + u^2) = (\sqrt{2} - 1)\sqrt{2} = 2 - \sqrt{2}, \qquad \xi = \sqrt{1 - \frac{1}{\sqrt{2}}},$$

$$\eta^2 = u^2 - \xi^2 = \sqrt{2} - 1 - 1 + \frac{1}{\sqrt{2}} = \frac{(\sqrt{2} - 1)^2}{\sqrt{2}}, \qquad \eta = \sqrt[4]{2} - \frac{1}{\sqrt[4]{2}}.$$

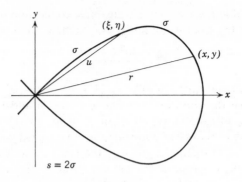

Figure 4

This point is easily constructed by means of compass and ruler; it is the mid-point of the full lemniscatic arc in the first quadrant. The length of the associated arc σ is thus one eighth of the length of the whole lemniscate.

We conclude with a comment which is not found in Fagnano but turns up in the later development of the theory of elliptic functions. Equation (12) can be verified directly by applying the substitution (16) to the lemniscatic integral $s(r)$. In our development, which follows Fagnano, this substitution resulted from the composition of the two simpler substitutions (8) and (10). In the corresponding integral relations (9) and (11) the integrals involving t are no longer lemniscatic. This blemish can be removed by dividing the auxiliary variable t by an eighth root of unity, namely,

$$\varepsilon = e^{\pi i/4} = \frac{1 + i}{\sqrt{2}}.$$

If we put $t = \varepsilon v$, then $t^2 = \varepsilon^2 v^2 = iv^2$, $t^4 = -v^4$ and

$$\sqrt{2}\, dt = \sqrt{2}\, \varepsilon\, dv = (1 + i)\, dv.$$

Now the relations (8) and (9) take the form

$$(19) \qquad r^2 = \frac{2iv^2}{1 - v^4}, \qquad \int_0^r \frac{dr}{\sqrt{1 - r^4}} = (1 + i) \int_0^v \frac{dv}{\sqrt{1 - v^4}}.$$

Similarly, if we divide (11) by ε and note that $\varepsilon^{-1}\sqrt{2} = 1 - i$, we obtain from (10), (11)

$$(20) \qquad v^2 = \frac{-2iu^2}{1 - u^4}, \qquad \int_0^v \frac{dv}{\sqrt{1 - v^4}} = (1 - i) \int_0^u \frac{du}{\sqrt{1 - u^4}}.$$

The integrals in (19) and (20) are all lemniscatic. The upper limit v appearing in two of these integrals is, of course, complex, and (19) and (20) yield formulas for the multiplication of the lemniscatic integral by the complex factors $1 + i$ and $1 - i$. Since $(1 + i)(1 - i) = 2$, composition of the relations (19) and (20) yields the earlier results on the doubling of a lemniscatic arc.

Complex multiplication of the lemniscatic integral and, more generally, of elliptic integrals, was first studied systematically in the 19th century, in particular by Abel and Kronecker. As we have just seen, however, a simple special case, namely, multiplication by $1 \pm i$, is already implied by the discovery of Fagnano.

2. The Euler addition theorem

The deeper significance of the doubling of the lemniscatic arc by Fagnano escaped his contemporaries for more than thirty years. It was not until 1751 that Euler began his investigations closely related to the work of Fagnano. After a few years these investigations led to the discovery by Euler of the addition theorem for elliptic integrals. In this section we shall attempt to reconstruct Euler's train of thought.

In the preceding section, in connection with the derivation of the relations obtained by Fagnano, we had occasion to introduce, as a sort of model, the substitution (7) which rationalizes the integral (6) for arc sin r. In this connection we see readily that there is a relation for the doubling of a circular arc which is quite analogous to that for the doubling of a lemniscatic arc; namely, if we put

$$r = 2u\sqrt{1 - u^2},$$

then we obtain

$$\int_0^r \frac{dr}{\sqrt{1 - r^2}} = 2\int_0^u \frac{du}{\sqrt{1 - u^2}}.$$

This result can also be obtained by direct algebraic computation or by using the transcendental substitution $u = \sin x$. This result includes the relation

$$\sin (2x) = 2 \sin x \cos x$$

which is a special case of the trigonometric addition theorem

$$\sin (x + y) = \sin x \cos y + \cos x \sin y.$$

The latter can be obtained by means of the substitutions

$$u = \sin x, \qquad v = \sin y, \qquad u\sqrt{1 - v^2} + v\sqrt{1 - u^2} = \sin(x + y)$$

and the formation of inverse functions and takes the form

$$(1) \qquad \int_0^u \frac{du}{\sqrt{1 - u^2}} + \int_0^v \frac{dv}{\sqrt{1 - v^2}} = \int_0^r \frac{dr}{\sqrt{1 - r^2}}, \qquad r = u\sqrt{1 - v^2} + v\sqrt{1 - u^2}.$$

To avoid difficulties resulting from the multiple-valued nature of the inverse trigonometric functions, u and v are supposed to be sufficiently small, non-negative numbers and the various square roots are taken to be positive.

Is there a corresponding version of the addition theorem (1) for the lemniscatic integral? In order to obtain an appropriate substitution we must invent a sensible generalization of the substitution

$$(2) \qquad\qquad r = \frac{2u\sqrt{1 - u^4}}{1 + u^4}$$

given by (16) in Section 1 which would, in a sense, reflect the transition from $r = 2u\sqrt{1 - u^2}$ to $r = u\sqrt{1 - v^2} + v\sqrt{1 - u^2}$. It is plausible to replace the numerator $2u\sqrt{1 - u^4}$ in (2) by the expression $u\sqrt{1 - v^4} + v\sqrt{1 - u^4}$. As denominator we choose the simplest symmetric function of u and v which becomes $1 + u^4$ for $u = v$, namely, $1 + u^2v^2$. We thus arrive at the substitution

$$(3) \qquad\qquad r = \frac{u\sqrt{1 - v^4} + v\sqrt{1 - u^4}}{1 + u^2v^2}.$$

In order to prove the corresponding addition theorem in the most convenient way, we view r in (3) as constant, u as an independent variable, and v as a dependent variable. For $u = 0$ we then have $v = r$. If we introduce the abbreviations

$$U = \sqrt{1 - u^4}, \qquad V = \sqrt{1 - v^4}$$

and differentiate (3) with respect to u, we obtain

$$(4) \quad \frac{du}{U}[(UV - 2vu^3)(1 + u^2v^2) - 2(uUV + v - vu^4)uv^2]$$

$$+ \frac{dv}{V}[(VU - 2uv^3)(1 + u^2v^2) - 2(vVU + u - uv^4)vu^2] = 0.$$

The two expressions in brackets in (4) have the common value $UV(1 - u^2v^2) - 2uv(u^2 + v^2)$ which becomes $\sqrt{1 - r^4}$ for $u = 0$, and is thus different from 0 when r and u are sufficiently close to 0. Hence v, as a function of u, satisfies the differential equation

$$\frac{du}{\sqrt{1 - u^4}} + \frac{dv}{\sqrt{1 - v^4}} = 0.$$

Integration with respect to u with lower limit 0 yields

$$\int_0^u \frac{du}{\sqrt{1 - u^4}} + \int_r^v \frac{dv}{\sqrt{1 - v^4}} = 0,$$

and thus the required addition theorem

$$s(u) + s(v) = s(r), \qquad r = \frac{u\sqrt{1 - v^4} + v\sqrt{1 - u^4}}{1 + u^2v^2},$$

which for $u = v$ reduces to the result of Fagnano. This was discovered by Euler in 1753. Obviously, Euler's result includes the assertion that two lemniscatic arcs given by their end points can be added by elementary geometric means.

Shortly after this discovery Euler took the last essential step towards a complete addition theorem for elliptic integrals. Namely, he replaced the expression $1 - u^4$ under the radical sign by the polynomial

$$P(u) = 1 + au^2 - u^4$$

with arbitrary constant a and generalized (3) accordingly to

$$(5) \qquad r = \frac{uV + vU}{1 + u^2v^2}, \qquad U = \sqrt{P(u)}, \qquad V = \sqrt{P(v)}.$$

Again taking r to be constant and differentiating we obtain in place of the expression in brackets the expressions

$$(UV + auv - 2vu^3)(1 + u^2v^2) - 2(uUV + v + avu^2 - vu^4)uv^2$$

and

$$(VU + avu - 2uv^3)(1 + u^2v^2) - 2(vVU + u + auv^2 - uv^4)vu^2,$$

which have the common value $(UV + auv)(1 - u^2v^2) - 2uv(u^2 + v^2)$. It follows that

$$\frac{du}{U} + \frac{dv}{V} = 0,$$

and this leads to the addition theorem

$$(6) \qquad \int_0^u \frac{du}{\sqrt{P(u)}} + \int_0^v \frac{dv}{\sqrt{P(v)}} = \int_0^r \frac{dr}{\sqrt{P(r)}},$$

$$r = \frac{u\sqrt{P(v)} + v\sqrt{P(u)}}{1 + u^2v^2}, \qquad P(u) = 1 + au^2 - u^4.$$

In view of the singularities of the integrand and the two-valued nature of the various square roots, it is necessary to investigate the domain of validity of these relations; this, however, we shall not do since we merely wished to sketch Euler's train of thought. It is easy to see that for a fixed a and small absolute values of the complex magnitudes u and v these relations hold provided all square roots refer to the same branch and the paths of integration are rectilinear.

Finally, Euler also investigated the general case when $P(u)$ is an arbitrary polynomial of degree four. Then, naturally, the computations become more complicated and we do not propose to go into the matter except for the following observation. If we subject the variable u to a fractional linear transformation

$$u = \frac{\alpha w + \beta}{\gamma w + \delta}$$

with complex parameters α, β, γ, δ and nonzero determinant $\alpha\delta - \beta\gamma \neq 0$, then the function

$$H(w) = (\gamma w + \delta)^4 P(u)$$

is again a polynomial in w of degree four whose five coefficients are homogeneous polynomials in α, β, γ, δ. Using simple algebraic arguments we can show that, in general, if we assign suitable values to the complex parameters α, β, γ, δ, four of the five coefficients of $H(w)$ will take on preassigned values. In particular, $H(w)$ can be made to take the form

$$H(w) = 1 + aw^2 - w^4$$

with suitable a, and, since

$$du = \frac{\alpha\delta - \beta\gamma}{(\gamma w + \delta)^2} dw,$$

it follows that the expression

$$(\alpha\delta - \beta\gamma)^{-1} \frac{du}{\sqrt{P(u)}} = \frac{dw}{\sqrt{H(w)}}$$

is precisely of the form considered in the preceding paragraph. This approach yields the general form of the Euler addition theorem.

Euler's penetrating proof continues to elicit admiration to this very day. What is not satisfying about Euler's proof is the fact that one cannot tell a priori that the substitution (3) will yield the required result. The addition theorems for the trigonometric functions become quite transparent when we bring in the connection with the exponential function, and it is reasonable to conjecture that the Euler addition theorem can also be proved more simply with the aid of function-theoretical arguments if we study thoroughly the relevant integrals as analytic functions in their full domains of definition. This will be our aim in this chapter.

At this point it is appropriate to give a general definition of elliptic integrals. Let $P(x)$ be a polynomial with complex coefficients of degree at most four, that is,

$$P(x) = a_0 x^4 + a_1 x^3 + a_2 x^2 + a_3 x + a_4.$$

Further, let $R(x, y)$ be a rational function of the two independent variables x and y with complex coefficients. We put

$$(7) \qquad\qquad y = \sqrt{P(x)}$$

and postpone the discussion of the two-valued nature of the square root until Section 5. We assume that the substitution (7) does not reduce the denominator of the rational function $R(x, y)$ to 0. Now we consider the integral

$$I(x) = \int_a^x R(x, y)\, dx,$$

where the integration is carried out over a definite path in the complex plane joining a to x. If the polynomial $P(x)$ has a multiple zero c, then we can factor out $(x - c)^2$ and extract a square root of this factor. This would leave under the square root sign a polynomial of degree at most two. In that case, as is well known, the integration can be carried out in an elementary manner, and the same is true, of course, if the coefficients a_0 and a_1 are both 0. In the remaining general case, that is, in the case when $P(x)$ is a polynomial of third or fourth degree with simple zeros, $I(x)$ is referred to as an *elliptic integral*. This name derives from the fact that the arc length of an ellipse in standard position expressed as a function of the abscissa is given by a particular integral of this kind.

The theory of elliptic integrals was first studied systematically by Legendre and is subsumed under the general theory of abelian integrals treated in Chapter 4. The elliptic integral which turns up in the Euler addition theorem has the special form

$$I(x) = \int_a^x \frac{dx}{y} = \int_a^x \frac{dx}{\sqrt{P(x)}},$$

and is called an *elliptic integral of the first kind;* this for reasons which will be made clear in the theory of abelian integrals. In preparation for the function-theoretical investigation of the analytic function $I(x)$, we introduce in the following sections the notion of the Riemann surface of an arbitrary multiple-valued analytic function and consider it, in particular, for $P(x)$. With a view to more distant applications, the following considerations are more general than at first necessary.

3. Analytic continuation

Consider a function defined in the neighborhood of a point a of the complex z-plane by means of a power series

$$(1) \qquad\qquad f(z) = a_0 + a_1(z - a) + \cdots$$

whose open circle of convergence $K(a)$ has radius $r(a)$. If s is a point of $K(a)$, then $f(z)$ can be expanded in a power series

(2) $$f(z) = b_0(s) + b_1(s)(z - s) + \cdots$$

of powers of $(z - s)$ which is obtained from (1) by expanding the powers

$$(z - a)^n = [(z - s) + (s - a)]^n \qquad (n = 1, 2, \ldots)$$

by the binomial theorem and arranging the series (1) in powers of $(z - s)$. The coefficients $b_0(s)$, $b_1(s)$, ... are obtained as convergent power series in $(s - a)$, and for the radius of convergence of the series (2) we have the inequality

$$r(s) \geq r(a) - |s - a| > 0.$$

We say that *the series (2) arises from the series (1) as a result of rearranging the series at s.* If

$$r(s) > r(a) - |s - a|,$$

then the circle of convergence $K(s)$ of (2) extends beyond the circle $K(a)$, and (2) yields an analytic continuation of the function defined by (1) in $K(a)$ into the part of $K(s)$ not covered by $K(a)$. The power series (1) and (2) yield the same functional values in the disk

$$|z - s| < r(a) - |s - a|;$$

this is so because (2) is the result of rearrangement of (1). This disk is contained in the intersection D of $K(a)$ and $K(s)$ and, since both power series converge in D, their values must coincide in D (Figure 5).

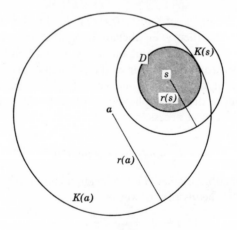

Figure 5

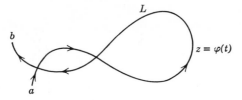

$z = \varphi(t)$

Figure 6

Let b be a point of the z-plane and let L be a curve joining a to b. As usual, a curve is determined by a complex-valued function $z = \varphi(t)$ defined and continuous on a real interval $\alpha \leq t \leq \beta$, and $\varphi(\alpha) = a$ and $\varphi(\beta) = b$ are called the end points of the curve. Such a parametric representation of the curve induces an ordering of its points and, in case of existence of multiple points, coinciding points are distinguished as corresponding to different values of the parameter t (Figure 6). If b and a coincide, then L is said to be closed. A closed curve is called a simple closed curve if the values of $\varphi(t)$ for t in the interval $\alpha \leq t < \beta$ are distinct.

If we divide the interval $\alpha \leq t \leq \beta$ into n subintervals by means of an increasing sequence of points $\alpha = \gamma_0, \gamma_1, \ldots, \gamma_n = \beta$, then L splits into n subarcs $L_1, L_2, \ldots, L_n$ with L_g ($g = 1, \ldots, n$) joining $\varphi(\gamma_{g-1}) = c_{g-1}$ to $\varphi(\gamma_g) = c_g$. Now let $K_0, K_1, \ldots, K_n$ be disks with centers $a = c_0, c_1, \ldots, c_n = b$, and let the closed arc L_g lie completely in K_{g-1} for $g = 1, \ldots, n$ (Figure 7). Further, let $f_0(z)$ be the power series (1) and, for $g = 1, \ldots, n$, let $f_g(z)$ which results from the rearrangement of $f_{g-1}(z)$ at c_g converge in K_g. Then $f_1, \ldots, f_n$ have been recursively defined and we say that the power series $f_n(z)$ is the result of *analytic continuation of the power series $f(z)$ along the curve L.*

We must now show that this analytic continuation is independent of the choice of division points on L. To this end we consider first the special case when the curve L is contained in K_0, and assert that the series $f_n(z)$ arises from $f_0(z)$ as a result of rearrangement at b. For $n = 1$ this assertion is trivially true. For $n > 1$ it follows from the induction assumption that f_{n-1} is the result of rearrangement of f_0 at c_{n-1}, so that the values $f_0(z)$ and $f_{n-1}(z)$ coincide in the intersection of K_0 and K_{n-1}. On the other hand, the values $f_{n-1}(z)$ and $f_n(z)$ coincide in the intersection of K_{n-1} and K_n. Since the point b is common to both intersections, our assertion follows in the case under consideration. In the general case, let δ be the least of the distances between L_g and the circumferences of the disks K_{g-1} ($g = 1, \ldots, n$), and let ρ be a positive number $< \delta$. In view of the continuity of $\varphi(t)$ it is possible, using suitable division points $a = d_0, d_1, \ldots, d_m = b$, to achieve so fine a subdivision of L into a sequence of arcs $T_1, T_2, \ldots, T_m$ that, for $h = 1, \ldots, m$, each point of T_h is contained in both disks G_{h-1} and G_h with centers d_{h-1} and d_h and common radius ρ (Figure 8). We are assuming that the old

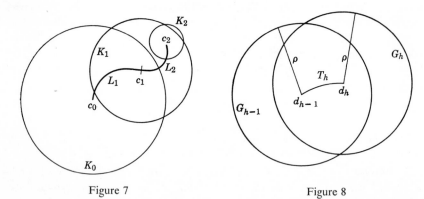

Figure 7   Figure 8

dividing points $c_0, \ldots, c_n$ are included among the new ones. In particular,
let $d_{h-1} = c_{g-1}$, $d_h, \ldots, d_j = c_g$ be the successive new division points
belonging to L_g. Since the power series f_{g-1} converges in K_{g-1} and all the disks
$G_{h-1}, G_h, \ldots, G_j$ lie in K_{g-1}, we conclude, on the basis of the part of our
assertion already established, that the series f_g is the analytic continuation of
f_{g-1} along L_g when we use the disks $G_h, \ldots, G_j$. If we make use of this for
$g = 1, \ldots, n$, we see that, for either subdivision, the analytic continuation
of $f(z)$ along L yields the same power series $f_n(z)$. Finally, consider the
analytic continuation of $f(z)$ for two subdivisions of L with values δ_1 and δ_2
of δ. By choosing $\rho < \min(\delta_1, \delta_2)$ we can construct a common refinement
of both subdivisions, and so establish the correctness of our assertion in the
general case.

In the preceding proof we used only the assumption that the arc T_h was
contained in the disk G_{h-1} for $h = 1, \ldots, m$. If we make the additional
assumption that T_h lies also in G_h, then it follows that analytic continuation
of $f_n(z)$ along the curve obtained by reversing the direction of the curve L
yields again the series $f_0(z)$. Further, since we can include any point c of L
among the division points $d_0, \ldots, d_m$, we see that the power series obtained
from $f(z)$ by continuation along the subarc of L with end point c can be
continued up to b and yields there the old power series $f_n(z)$.

In defining analytic continuation along L the basic assumption was the
existence of a chain of disks $K_0, K_1, \ldots, K_n$ such that: (1) the centers $a =$
$c_0, c_1, \ldots, c_n = b$ of these disks are successive points of L, (2) each of the
subarcs L_g $(g = 1, \ldots, n)$ is contained in K_{g-1}, and (3) for $g = 1, \ldots, n$ the
power series f_g resulting from the rearrangement of f_{g-1} at c_g converges in K_g.
If no such chain of disks exists, then we say that the power series $f(z)$ cannot
be continued analytically along L. This case merits additional attention.
Thus let a point c traverse the curve L monotonically from a to b and let
$L(c)$ be the subarc of L from a to c. If $L(c)$ lies entirely in K_0, then $f(z)$ can

certainly be continued analytically along $L(c)$. This implies the existence of a point d on L which precedes b and has the property that $f(z)$ can be continued analytically along every $L(c)$ with c preceding d but not along $L(d)$. In this connection it should be pointed out that there exists no general constructive procedure for deciding in a specific case whether or not it is possible to continue $f(z)$ analytically along all of L. In the case of really interesting functions the issue is invariably settled by making use of their special properties.

In investigating the possibility of continuation of the function $f(z)$ from a to b it is essential that the curve L be given. If L^* is another curve joining a to b, then it is quite possible that the function can be continued along one of these curves all the way to the end point b but not along the other curve. It is also possible that the function can be continued analytically along both curves but that the power series obtained at b are different. The following result is true, however. If the given function $f(z)$ can be continued analytically along L, then it can also be continued analytically along every sufficiently close curve L^* and the result is the same power series at the end point b. To make the notion of sufficient closeness precise we assume that L and L^* have representations $z = \varphi(t), z^* = \varphi^*(t), \alpha \le t \le \beta$, involving the same parameter t, and that the distance $|z - z^*|$ between corresponding points on the two curves is sufficiently small, namely, that it does not exceed half the magnitude δ introduced above. We now choose $\rho \le \delta/2$ and again decompose L into the subarcs T_h ($h = 1, \ldots, m$) with initial points d_{h-1}. We then obtain a corresponding decomposition of L^* into subarcs T_h^* with initial points d_{h-1}^*. Here T_h and T_h^* are contained entirely in the disk

$$(3) \qquad\qquad |z - d_{h-1}| < \delta,$$

and the power series resulting from analytic continuation of $f(z)$ along L up to d_{h-1} converges in that disk. Suppose now that we have already proved that it is possible to continue the given power series $f(z)$ at a analytically along L^* up to d_{h-1}^*, and that the resulting power series coincides with the power series obtained by continuation of $f(z)$ along L up to d_{h-1} and from there along the straight line segment joining d_{h-1} and d_{h-1}^*. This is again trivial for $h = 1$. Since T_h and T_h^* lie in the disk (3), analytic continuation of the series obtained at d_{h-1}^* along T_h^* yields the same result which we would obtain by going first along the segment from d_{h-1}^* to d_{h-1}, then along T_h and, finally, along the segment from d_h to d_h^*. This induction argument yields the required result for $h = m$.

Since every curve can be approximated with arbitrary accuracy by means of a polygonal curve with vertices on the curve, it follows that, when investigating analytic continuation, we may assume that the curve L is a polygonal curve, or, at least, that it is piecewise smooth. This will occasionally prove to be useful.

The different power series obtained by means of all possible analytic continuations to arbitrary points are called *function elements* of the analytic function determined by the power series (1).

4. Riemann regions

A complex domain is an open connected set in the z-plane. The example of the function $w = z^2$ in the punctured plane shows that conformal mappings effected by very simple analytic functions need not, in general, be one-one. Thus in studying functions of a complex variable it is useful to introduce a suitable generalization of the notion of a domain known as a *Riemann region.**

We avoid for the time being the notion of a branch point and define a Riemann region $\mathfrak{G}$ over the z-plane as follows: Every point $\mathfrak{z}$ of $\mathfrak{G}$ has a complex coordinate z and $\mathfrak{z}$ is said to *lie above its projection* z. If $\mathfrak{b}$ is an arbitrary point of $\mathfrak{G}$ and b is its coordinate, then we associate with $\mathfrak{b}$ on $\mathfrak{G}$ a definite neighborhood $K_{\mathfrak{b}}$ which lies above the circular domain

$$|z - b| < r_{\mathfrak{b}}$$

whose radius $r_{\mathfrak{b}}$ depends on $\mathfrak{b}$ in an as yet unspecified manner. Over every point z of this circular domain K_b there is supposed to lie exactly one point $\mathfrak{z}$ of $K_{\mathfrak{b}}$. Thus $K_{\mathfrak{b}}$ is to be thought of as a disk above the z-plane precisely covering K_b. We call $K_{\mathfrak{b}}$ a *surface element* with center $\mathfrak{b}$. Now let $\mathfrak{c} \neq \mathfrak{b}$ be a point of $\mathfrak{G}$ over c. We stipulate that the intersection $K_{\mathfrak{b}} \cap K_{\mathfrak{c}}$ of the two surface elements $K_{\mathfrak{b}}$ and $K_{\mathfrak{c}}$ is empty or, otherwise, $K_{\mathfrak{b}}$ and $K_{\mathfrak{c}}$ share precisely those points $\mathfrak{z}$ which lie over the points z of the intersection $K_b \cap K_c$ of K_b and K_c. In the latter case we say that $K_{\mathfrak{b}}$ and $K_{\mathfrak{c}}$ are *linked*. It can therefore happen that the circular domains K_b and K_c overlap and the surface elements $K_{\mathfrak{b}}$ and $K_{\mathfrak{c}}$ are disjoint, as, for example, in case when $b = c$ and $\mathfrak{b} \neq \mathfrak{c}$.

It is easy to show that if $K_a \cap K_b \cap K_c$ is not empty and the pairs $K_{\mathfrak{a}}$, $K_{\mathfrak{b}}$ and $K_{\mathfrak{b}}$, $K_{\mathfrak{c}}$ are linked, then the pair $K_{\mathfrak{a}}$, $K_{\mathfrak{c}}$ is also linked.

It is not difficult to see how to define a curve on $\mathfrak{G}$. By a curve on a surface element $K_{\mathfrak{b}}$ we mean the image in $K_{\mathfrak{b}}$ of a curve on the disk K_b under the one-one correspondence between K_b and $K_{\mathfrak{b}}$; we call such a curve a *curve element*. Now a *curve* $\mathfrak{L}$ on $\mathfrak{G}$ is determined by a function

$$\mathfrak{z} = \mathfrak{y}(t)$$

from a real interval $\alpha \leq t \leq \beta$ to $\mathfrak{G}$ such that for every t_1 in that interval there exists a subinterval

$$t_1 - \varepsilon \leq t \leq t_1 + \varepsilon$$

---

* Sometimes referred to as a *Riemann domain*.

with the property that the points $\eta(t)$ with t in that subinterval form a curve element. One can show in the usual way that $\mathfrak{L}$ can be covered by means of a finite number of curve elements. Thus every curve $\mathfrak{L}$ in $\mathfrak{G}$ issuing from a point $\mathfrak{b}$ lies over a projected curve L in the z-plane issuing from the projection b of $\mathfrak{b}$. The curve $\mathfrak{L}$ on $\mathfrak{G}$ is completely determined by its initial point $\mathfrak{b}$ and its projection L. Here it is essential that $\mathfrak{G}$, as defined, has, as yet, no branch points. A final assumption concerning $\mathfrak{G}$ is that it is connected, that is, that any two of its points $\mathfrak{b}$ and $\mathfrak{c}$ can be joined by means of a curve on $\mathfrak{G}$. Of course, $\mathfrak{G}$ is supposed to be nonempty. It must be borne in mind that a Riemann region can be put together out of surface elements in different ways. Thus, for example, in defining the $K_{\mathfrak{b}}$ and the K_{b} one could replace the radii $r_{\mathfrak{b}}$ with some smaller values. As a concrete example, we could cover the complex plane with the family of concentric disks $|z| < n$ $(n = 1, 2, \ldots)$, and again, with the family of disks $|z - m - ni| < 1$ $(m, n = 0, \pm 1, \ldots)$.

It is clear what one means by a function $f(\mathfrak{z})$ on $\mathfrak{G}$, namely, in every neighborhood $K_{\mathfrak{b}}$, $f(\mathfrak{z})$ stands for a function $f_{\mathfrak{b}}(z)$, in the usual sense of the word, defined in $|z - b| < r_{\mathfrak{b}}$. This definition implies that $f(\mathfrak{z})$ is single-valued on $\mathfrak{G}$. A function f is called *regular* on $\mathfrak{G}$ if for each point $\mathfrak{b}$ the function $f_{\mathfrak{b}}(z)$ can be represented by a power series in the local parameter $t = z - b$ which converges in the corresponding disk $K_{\mathfrak{b}}$; in this case the power series $f_{\mathfrak{b}}(z)$ is called the *function element* of f in $K_{\mathfrak{b}}$. Now let $\mathfrak{a}$ be a fixed point on $\mathfrak{G}$ over a and let $\mathfrak{L}_1$, and $\mathfrak{L}_2$ be two curves on $\mathfrak{G}$ which join $\mathfrak{a}$ to the same point $\mathfrak{b}$ and have projections L_1 and L_2. If

$$f_{\mathfrak{a}}(z) = f_0(z) = a_0 + a_1(z - a) + \cdots$$

is the function element of $f(\mathfrak{z})$ in $K_{\mathfrak{a}}$, then analytic continuation of $f_0(z)$ along L_1 and along L_2 leads to the same function element $f_{\mathfrak{b}}(z)$.

Now we reverse the order of our considerations and start with a given power series $f_0(z)$ in the variable $(z - a)$. We pose the following question: Does there exist a Riemann region $\mathfrak{G}$ and a regular function $f(\mathfrak{z})$ on $\mathfrak{G}$ such that the totality of analytic continuations of $f_0(z)$ along curves in the z-plane yield precisely the totality of function elements of f? To construct $\mathfrak{G}$ and f, we consider for an arbitrary point b in the z-plane all power series at b, if any, resulting from analytic continuation of the given power series along any path joining a to b. With each of these power series we associate in a one-one manner a point $\mathfrak{b}$ over b and a surface element $K_{\mathfrak{b}}$ which lies precisely over the circle of convergence K_b of the power series in question; $K_{\mathfrak{b}}$ is obtained by associating to each point c in K_b the point $\mathfrak{c}$ associated with the rearrangement of $f_{\mathfrak{b}}(z)$ at c. (We observe that for a fixed b and two different points $\mathfrak{b}$ the two associated surface elements $K_{\mathfrak{b}}$ are entirely separated; at the same time it is not, in general, expedient to form an idea of the spatial arrangement of the surface elements over b.) We do this for all points b which can at all be reached by analytic continuation of $f(z)$ from a. In particular, we assign a point $\mathfrak{a}$ over a to the power series $f_0(z)$. We now come to the principal issue, namely, the rule for linking two surface elements $K_\mathfrak{b}$ and $K_\mathfrak{c}$. We assume that $\mathfrak{b}$ is over b, $\mathfrak{c}$ is over c and $\mathfrak{b} \neq \mathfrak{c}$. If the projections K_b and K_c of $K_\mathfrak{b}$ and $K_\mathfrak{c}$ on the z-plane are disjoint, then, of course, $K_\mathfrak{b}$ and $K_\mathfrak{c}$ must be regarded as entirely separated. If the intersection $D = K_b \cap K_c$ is not empty, then we consider the power series $f_b(z)$ and $f_c(z)$ resulting from analytic continuation of $f_0(z)$ which are paired off with the surface elements $K_\mathfrak{b}$ and $K_\mathfrak{c}$. Rearrangement of the power series $f_b(z)$ and $f_c(z)$ at a point d of D yields two power series in the variable $(z - d)$ which may or may not be identical; this independently of the choice of d in D. In the first case $f_c(z)$ is obtained from $f_b(z)$ by means of analytic continuation along the segment from b to c and, in that case, we identify precisely those points of $K_\mathfrak{b}$ and $K_\mathfrak{c}$ which lie over D; this is the case when $K_\mathfrak{b}$ and $K_\mathfrak{c}$ are linked. In the second case $K_\mathfrak{b}$ and $K_\mathfrak{c}$ are not linked.

Linking the various surface elements in the manner just described yields a Riemann region $\mathfrak{G}$ and a regular single-valued function $f(\mathfrak{z})$ in $\mathfrak{G}$ which is defined by each function element $f_b(z)$. This answers the question posed above. A related question is that of the existence of another Riemann region $\mathfrak{G}'$ with the same properties as $\mathfrak{G}$. It can be shown that any such Riemann region $\mathfrak{G}'$ is essentially the same as $\mathfrak{G}$. Thus there is essentially only one Riemann region $\mathfrak{G}$ with the desired properties.

We shall obtain the *Riemann surface* of the function $f(\mathfrak{z})$ by adding to $\mathfrak{G}$ certain additional points, namely, *branch points** and *points at infinity;* thus $\mathfrak{G}$ has only provisional character and will be called the *punctured Riemann surface* of the given function. In obtaining all function elements by analytic continuation we could start with an arbitrary $f_b(z)$ in place of $f_a(z)$. It is thus clear that the punctured Riemann surface does not depend on the initial point $\mathfrak{a}$.

We come now to the definition of branch points and consider first Riemann regions which meet the following special requirement: Let $\dot{K}$ be the punctured disk $0 < |z - c| < r$, let b be a point of $\dot{K}$, and let $\mathfrak{b}$ be a point in $\mathfrak{G}$ over b. If L is any curve in $\dot{K}$ issuing from b, then we require the existence on $\mathfrak{G}$ of a curve $\mathfrak{L}$ issuing from $\mathfrak{b}$ and lying over L.

In particular, let L be a piecewise smooth closed curve in $\dot{K}$. Since L issues from b and is closed, it terminates at b. While the end point $\mathfrak{b}_L$ of $\mathfrak{L}$ need not coincide with $\mathfrak{b}$, it remains unchanged when the point $\mathfrak{b}$ is kept fixed and L is deformed continuously while confined to $\dot{K}$. This is established in much the same way as the assertion that analytic continuation of a function element along two sufficiently close curves with common end point yields the

---

* Also referred to as *ramification points*. The two terms will be used interchangeably. (Tr.)

same function element (cf. preceding section). We call the integer

$$v(L) = \frac{1}{2\pi i} \int_L \frac{dz}{z - c} = m$$

the *winding number* of L. This integral gives the change of $\arg(z - c)$ in multiples of 2π when L is traversed, and it obviously remains unchanged when the piecewise smooth curve L is continuously deformed. Now let m be a preassigned integer, and let L_m be the special curve obtained by going from b along a radius of $\dot{K}$, coming close to the center c, circling c m times, and returning along the same radius to b; here the circular path is traversed in a positive or negative direction depending on the sign of m. It can be shown that it is possible to shrink L continuously in $\dot{K}$ to L_m while keeping b fixed. Thus the point $\mathfrak{b}_L$ depends only on the winding number $v(L) = m$ and we can put $\mathfrak{b}_L = \mathfrak{b}_m$ $(m = 0, \pm 1, \ldots)$. It can happen that all the points $\mathfrak{b}_m$ are distinct as, for example, in the case of the Riemann surface of log $(z - c)$. We are concerned, however, with the case when $\mathfrak{b}_k = \mathfrak{b}_l$ for two distinct indices k and l, where we can assume that $k - l = j > 0$. If z traverses the curve L_{k-l}, then it follows that $\mathfrak{b}_{m+j} = \mathfrak{b}_m$ $(m = 0, \pm 1, \ldots)$, so that $\mathfrak{b}_m$ has period j in m. If j is given its least value, then the j points $\mathfrak{b} = \mathfrak{b}_0, \mathfrak{b}_1, \ldots, \mathfrak{b}_{j-1}$, are distinct and $\mathfrak{b}_j = \mathfrak{b}$.

Now let Q be an arbitrary curve in $\dot{K}$ issuing from b and terminating at z, and let $\mathfrak{Q}$ be the curve in $\mathfrak{G}$ which issues from $\mathfrak{b}$, lies over Q and terminates at $\mathfrak{z}$. All such points $\mathfrak{z}$ form a Riemann region $\mathfrak{G}^*$ in $\mathfrak{G}$ lying over $\dot{K}$. We are about to investigate $\mathfrak{G}^*$ in some detail. It is clear that every closed curve M in $\dot{K}$ issuing from z can be continuously deformed to a curve obtained by first traversing the curve Q from z to b (that is, in opposite direction), and then some curve L from b to b, and, finally, the curve Q from b to z. It follows that, in traversing the curve on $\mathfrak{G}^*$ which issues at $\mathfrak{z}$ and lies over M, we return to $\mathfrak{z}$ if and only if the winding number $v(M)$ is divisible by j. Now put

$$z - c = t^j, \qquad \sqrt[j]{b - c} = t_0, \qquad \sqrt[j]{r} = s > 0,$$

where t_0 is a definite one of the j distinct values of the root. This defines a conformal map of the punctured disk $\dot{G}$ in the t-plane defined by $0 < |t| < s$ onto the domain $\dot{K}$ covered j times; in particular, t_0 is mapped onto b. Now let A be a curve in $\dot{G}$ issuing from t_0 and terminating at t and let B be its image in $\dot{K}$. Consider first the case when $t_0 = t$, that is, when A is closed. Since

$$\frac{dz}{z - c} = j\frac{dt}{t},$$

it follows that the winding number of B is a multiple of j and the curve on $\mathfrak{G}^*$ which lies over B leads from $\mathfrak{b}$ to $\mathfrak{b}$. Now if A_1 and A_2 are two curves in

$\dot{G}$ which go from t_0 to t, then A_1 followed by the curve obtained by reversing the direction of A_2 yield a closed curve A. Thus, if B_1 and B_2 are the images of A_1 and A_2 in $\dot{K}$ leading from b to z, and if, beginning at $\mathfrak{b}$, we traverse the curves in $\mathfrak{G}^*$ lying over B_1 and B_2, then in each case we arrive at the same point $\mathfrak{z}$ over z.

Now we introduce on $\mathfrak{G}^*$ in place of z the new coordinate t. Specifically, the new coordinate of the point $\mathfrak{b}$ over b is t_0 and the new coordinate of an arbitrary $\mathfrak{z}$ on $\mathfrak{G}^*$ is $t = \sqrt[j]{z - c}$. To determine the value of the root we join $\mathfrak{b}$ to $\mathfrak{z}$ by means of a curve $\mathfrak{B}$ lying over the curve B joining b to z. Then B is again the image of a definite curve A issuing from t_0 whose end point is precisely t. The value t depends solely on $\mathfrak{z}$ and not on the choice of the curve B. Further, there are exactly j distinct points $\mathfrak{z}$ on $\mathfrak{G}^*$ lying over z and their t-coordinates are the j values of the root. A concrete representation of $\mathfrak{G}^*$ is obtained in the following familiar manner: We introduce in $\dot{G}$ polar coordinates by means of the substitution

$$t = pe^{(2\pi i/j)u}$$

with $0 < p < s$ and $0 \le u < j$. The prescription $h - 1 \le u < h$ ($h = 1, \dots, j$) divides $\dot{G}$ into j sectors of angular width $2\pi/j$. The mapping $z = c + t^j$ carries each sector $\dot{G}_h$ into the punctured, cut disk $\dot{K}_h$ obtained by cutting $\dot{K}$ along the radius from c to $c + r$. The edges of the cuts of the $\dot{K}_h$ are linked in the manner of the edges of the cyclically arranged $\dot{G}_h$, that is, the lower edge of $\dot{K}_h$ is identified with the upper edge of $\dot{K}_{h+1}$ ($h = 1, \dots, j - 1$) and the lower edge of $\dot{K}_j$, with the upper edge of $\dot{K}_1$. The resulting j-sheeted punctured *ramified surface* is regarded as a new surface element of $\mathfrak{G}$, and this addition yields $\mathfrak{G}^*$. In the case $j = 1$, $\mathfrak{G}^*$ results from taking $\dot{K}$ itself as a new surface element of $\mathfrak{G}$.

In particular, let $\mathfrak{G}$ be the punctured Riemann surface of a regular function $f(\mathfrak{z})$, and let

(1) $$f_\mathfrak{b}(z) = b_0 + b_1(z - b) + \cdots$$

be the function element at the point $\mathfrak{b}$ in $\mathfrak{G}^*$. As a result of the substitution $z = c + t^j$, $f_\mathfrak{b}(z) = g(t)$ becomes a function of t which is regular in a sufficiently small disk $|t - t_0| < q$. Since $f(\mathfrak{z})$ is single-valued on $\mathfrak{G}$, $g(t)$ can be continued analytically from t_0 to every point of the punctured disk $\dot{G}$ and this continuation is unique. It follows that $g(t)$ admits the following Laurent expansion in $\dot{G}$,

(2) $$g(t) = \sum_{n=-\infty}^{\infty} c_n t^n \qquad (0 < |t| < s),$$

and thus

(3) $$f_\mathfrak{b}(z) = \sum_{n=-\infty}^{\infty} c_n (z - c)^{n/j},$$

provided that the point z lies in the disk

$$|z - b| < \min (|b - c|, r - |b - c|),$$

and that the branch of the root

$$\sqrt[j]{z - c} = t_0\left(1 + \frac{z - b}{b - c}\right)^{1/j}$$

is uniquely determined there by means of the binomial series. Analytic continuation of $f_b(z)$ from b along a closed curve L in $\dot{K}$ with winding number m yields the function element

$$f_m(z) = f_{b_m}(z) \qquad (m = 0, \pm 1, \ldots),$$

while $\sqrt[j]{z - c}$ is multiplied by the factor ω^m with $\omega = e^{2\pi i/j}$. Thus the series (3) which converges in $\dot{K}$ yields for the j different values of the roots precisely the j branches $f_0(z), \ldots, f_{j-1}(z)$ which are obtained from the power series (1) as a result of arbitrary continuation in $\dot{K}$.

There arise three possibilities. First, all coefficients c_n with negative index n may vanish; in this case, the function represented by (2) is regular at $t = 0$, and the function $f(\mathfrak{z})$ given in the neighborhood of b by (3) approaches the definite limit c_o when z approaches c along a curve issuing from b and lying in the disk $|z - c| < r$. Second, there may exist a negative index p such that $c_p \neq 0$ and all $c_n = 0$ for $n < p$; in this case, $g(t)$ has a pole of order $-p$ at $t = 0$, and so $f(\mathfrak{z}) \to \infty$ for $z \to c$. Third, we may have the case when $c_n \neq 0$ for infinitely many negative indices n; in this case, as is well known, the function $g(t)$ comes arbitrarily close to every value in every neighborhood of the origin and a corresponding statement holds for $f(\mathfrak{z})$ at $z = c$. In the first two cases we make the punctured disk $\dot{G}$ into a full disk G by addition of the center $t = 0$ and, similarly, we complete the j-sheeted, punctured ramified surface $\mathfrak{G}^*$ by addition of the *branch point* $\mathfrak{c}$ over c. The completed ramified surface is designated as the surface element $K_{\mathfrak{c}}$; it arises, as we see, from the disk $|t| < s$ via the mapping $z = c + t^j$. Since the coordinate t is single-valued on the new surface element, we call it also the *local uniformizing parameter* or, simply, the *local parameter*. Thus in $K_{\mathfrak{c}}$, f should be regarded as a function of the parameter t with $\mathfrak{c}$ a regular point in the first case and a pole of order $-p$ in the second case. When $j = 1$, $f(\mathfrak{z})$ is already single-valued as a function of $\mathfrak{z}$ on $\mathfrak{G}^*$. Then $z = c$ cannot be a regular point for, otherwise, $\mathfrak{c}$ and the surface element $K_{\mathfrak{c}}$ would already belong to $\mathfrak{G}$.

Finally, it is necessary to extend the above considerations in a reasonable way to the case when, instead of c, we have the point at infinity. This requires only minor modifications. Thus we designate by $\dot{K}$ the circle exterior $r < |z| < \infty$ and we put

$$z = t^{-j}, \qquad \sqrt[j]{b} = t_0^{-1}, \qquad \sqrt[j]{r} = s^{-1}.$$

These are the only modifications needed, except that when $j = 1$ a new point is added to the punctured Riemann surface (even in the case of regularity since the punctured Riemann surface contains no point at infinity). To avoid the exceptional position of the point at infinity in the z-plane, we can use the stereographic projection to replace the z-plane with the complex z-sphere.

Our procedure shows how to obtain the Riemann surface $\mathfrak{R}$ of a function $f(\mathfrak{z})$ given locally by a power series $f_0(z)$ by adding to its punctured Riemann surface various new points and surface elements. The function $f(\mathfrak{z})$ is single-valued throughout $\mathfrak{R}$. In the neighborhood of every point it is a function of the local parameter t and as such it is representable by a power series in t with at most finitely many negative exponents.

An analog of the transition from the punctured to the full Riemann surface is the extension of a Riemann region $\mathfrak{G}$ by the addition of certain ramification points† and points over ∞. In this case, however, no function is usually apparent which has $\mathfrak{G}$ as its punctured Riemann surface. Consequently, it is necessary to state for each punctured ramified surface $\mathfrak{G}^*$ on $\mathfrak{G}$ whether or not the ramification point $\mathfrak{c}$ which lies over $z = c$ should be added to the new surface element $K_\mathfrak{c}$. This extension process yields a Riemann region $\mathfrak{R}$ which is a covering surface of the complex z-sphere and has an arbitrary system of branch points. It is natural to ask whether every Riemann region is also a Riemann surface. The answer to this question is in the affirmative, but the construction of an analytic function f, whose Riemann surface is precisely the preassigned Riemann region, requires lengthy preparations and will be accomplished only in Section 8 of Chapter 2.

Consider a function f given on a surface element $K_\mathfrak{c}$ of a Riemann region and suppose that, viewed as a function of the local parameter, f is regular at all points of $K_\mathfrak{c}$ except, possibly, for a pole at $\mathfrak{c}$. Then it may be possible or impossible to continue f analytically (admitting poles) in a unique manner along every path on $\mathfrak{R}$ issuing from $\mathfrak{c}$. In the first case, $f(\mathfrak{z})$ is well defined by analytic continuation on all of $\mathfrak{R}$ and, as a function of each particular local parameter, it is invariably regular to within poles. Such a function is said to be *analytic*‡ in $\mathfrak{G}$. Of course, by a *regular analytic* function on $\mathfrak{G}$ we mean a function which is analytic on $\mathfrak{G}$ and everywhere finite. It is obvious that the function z and, more generally, every rational function of z is meromorphic on every Riemann region.

5. The Riemann surface of the function
$$\sqrt{a_0 z^4 + a_1 z^3 + a_2 z^2 + a_3 z + a_4}$$

In the preceding section we showed in great detail how to construct a Riemann surface out of disk-like surface elements by means of suitable

---

† Cf. footnote 1.
‡ The word *analytic* will be used to denote functions which are possibly multi-valued. (Tr.)

identifications. For particular functions it is possible to simplify this construction by making use of larger components which are easily surveyed. We leave it to the reader to prove that the Riemann surfaces which we are about to construct are in fact the desired Riemann surfaces in the sense of our original definition.

Let

$$P(z) = a_0 z^4 + a_1 z^3 + a_2 z^2 + a_3 z + a_4$$

and let the analytic function $w = f(z)$ be defined by the equation $w^2 = P(z)$. We rule out the trivial case $a_0 = a_1 = a_2 = a_3 = 0$, for in that case w would be constant. We may assume that $P(z)$ has simple zeros [if c is a multiple zero of $P(z)$ then $(z - c)^{-1}w$ and w have the same Riemann surface], and that $P(z)$ has leading coefficient 1. This means that we need only consider the four possibilities:

$$P(z) = z - a, \qquad P(z) = (z - a)(z - b),$$

$$P(z) = (z - a)(z - b)(z - c), \qquad P(z) = (z - a)(z - b)(z - c)(z - d),$$

where a, b, c, d are four distinct complex numbers.

In the first case $w = \sqrt{z - a}$. We put two copies of the z-plane together and make in each of these sheets a simple cut L from a to ∞. We denote the edges of the cut in one sheet by L_1 and L_2, and the corresponding edges of the cut in the other sheet by L_2 and L_1. Now we join the four edges crosswise, that is we identify the edge L_k, $k = 1, 2$, in one sheet with the edge L_k in the other sheet. If we disregard for the moment the points a and ∞ in both sheets, we obtain a Riemann region $\mathfrak{G}$ in which the function w, which is two-valued in the z-plane, is single-valued. (This is the special case $j = 2$ of the general situation investigated in the preceding section.) Now $z \to a$ implies $w \to 0$ and $z \to \infty$ implies $w \to \infty$. It follows that the branch points 0 and ∞ must be added to $\mathfrak{G}$ as ordinary points. In this way we obtain the complete Riemann surface $\mathfrak{R}_1$ of the function $w = \sqrt{z - a}$. It should be pointed out that the nature of the simple cut from a to ∞ has no effect on the construction of $\mathfrak{R}_1$. Specifically, L need not be rectilinear; on the other hand, if L is chosen to be a ray from a to ∞, the direction of the ray is of no consequence.

In the second case, $w = \sqrt{(z - a)(z - b)}$. We could reduce this case to the previous case by using the fractional linear substitution $z - b = u^{-1}$ which carries the zero b into ∞ and reduces our function to the function

$$w = u^{-1}\sqrt{b - a}\sqrt{u + (b - a)^{-1}}.$$

With a view to future applications, however, it is preferable to investigate the case $w = \sqrt{(z - a)(z - b)}$ directly. To this end we repeat the construction carried out above except that now the simple curve L joins the points a

and b. As before, we cut two copies of the z-plane along L and join the resulting sheets $\mathfrak{B}_1$ and $\mathfrak{B}_2$ crosswise along the edges of the cuts (Figure 9). We disregard for the moment the points a, b, ∞, and show that the function w is single-valued on the resulting Riemann region $\mathfrak{G}$. To this end we consider the function w in the complex plane z cut along L where the edges L_1 and L_2 of L are regarded as distinct. If p is a point different from a and b then, using the binomial formula, we can expand each of the expressions

$$\sqrt{z-a} = \sqrt{p-a}\left(1 + \frac{z-p}{p-a}\right)^{\frac{1}{2}}, \qquad \sqrt{z-b} = \sqrt{p-b}\left(1 + \frac{z-p}{p-b}\right)^{\frac{1}{2}}$$

in some neighborhood of p in a power series in $z - p$. The same is true of w which is the product of these expressions. Since these expressions are two-valued, we obtain two power series w_1 and $w_2 = -w_1 \neq w_1$. Beginning at p we continue w_1 analytically in the cut z-plane $\mathfrak{B}_1$; here it is assumed that p is not a point of L_1 or L_2. Now let C be a path on $\mathfrak{B}_1$ leading from p back to p. The winding number of C relative to a is the same as its winding number relative to b; this follows from the fact that C does not meet the cut L and, therefore, its winding number is the same for an arbitrary point q on L. We conclude that traversing the path C results in multiplication of $\sqrt{z-a}$ and $\sqrt{z-b}$ by the same power of -1, and this implies that w_1 is single-valued on $\mathfrak{B}_1$. Similarly, w_2 is seen to be single-valued on $\mathfrak{B}_2$.

Now we consider on $\mathfrak{B}_1$ two corresponding points $\mathfrak{y}_1$ and $\mathfrak{y}_2$ of the edges L_1 and L_2. We connect these points by means of a curve D which issues from $\mathfrak{y}_1$, follows L_1 without reaching a, circles a once on $\mathfrak{B}_1$, follows L_2 and terminates at $\mathfrak{y}_2$. Since the winding numbers of D relative to a and b are ± 1 and 0 respectively, it follows that the function w_1, which is single-valued on $\mathfrak{B}_1$, takes on opposite values at the points $\mathfrak{y}_1$ and $\mathfrak{y}_2$. A similar statement holds for w_2 on $\mathfrak{B}_2$ (Figure 10). Cross-connection of the edges of $\mathfrak{B}_1$ and $\mathfrak{B}_2$ results in the identification of points carrying the same functional values. This implies that w is single-valued on $\mathfrak{G}$. Now, $z \to a$ or $z \to b$ implies

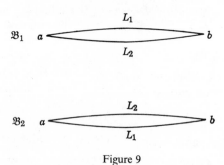

Figure 9

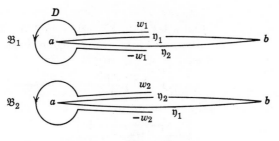

Figure 10

$w \to 0$; this means that a and b must be added to $\mathfrak{G}$ as branch points. Finally, to understand the behavior of our function at infinity, we put $z = t^{-1}$ and obtain

$$w = \pm t^{-1}(1 - at)^{\frac{1}{2}} (1 - bt)^{\frac{1}{2}}.$$

This shows that the points at infinity on $\mathfrak{B}_1$ and $\mathfrak{B}_2$ are simple poles of w_1 and 7_2 respectively. By adding the two points at infinity to $\mathfrak{G}$ we obtain the Riemann surface $\mathfrak{R}_2$ of $w = \sqrt{(z - a)(z - b)}$.

If we shift by stereographic projection from the z-plane to the z-sphere, then we see that the Riemann surface $\mathfrak{R}_1$ is obtained from $\mathfrak{R}_2$ by shifting the branch point b to ∞. Since the third and fourth cases are similarly related, it suffices to study one of them, say, the fourth case. Let

$$w = \sqrt{(z - a)(z - b)(z - c)(z - d)}.$$

We join a to b by means of a curve L, and c to d by means of a curve M disjoint from L. We then take two copies $\mathfrak{B}_1$ and $\mathfrak{B}_2$ of the z-plane cut along L and M and connect them crosswise along the edges of the cuts L and M (Figure 11). As before we show that both branches of each of the functions $\sqrt{(z - a)(z - b)}$ and $\sqrt{(z - c)(z - d)}$ are single-valued on $\mathfrak{B}_1$

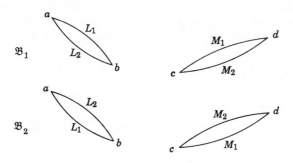

Figure 11

and $\mathfrak{B}_2$; this implies the single-valuedness of the function w itself. Again, arguing as before, we show that w takes on opposite values at corresponding edge points of the cuts L and M; this implies the single-valuedness of w on the cross-connected two-sheeted Riemann region. Next we add the four branch points a, b, c, d and the two points at infinity which in the present case are double poles of our function. In order to obtain from the Riemann surface $\mathfrak{R}_4$ just constructed the Riemann surface $\mathfrak{R}_3$ of

$$w = \sqrt{(z-a)(z-b)(z-c)},$$

we pass to the limit $d \to \infty$. This implies that $\mathfrak{R}_3$ has the four branch points a, b, c, ∞.

It is now easy to construct the Riemann surface of $w = \sqrt{P(z)}$, where $P(z)$ is a polynomial of degree n with n distinct zeros $c_1, \ldots, c_n$. Specifically, if n is even, $n = 2\nu$, then we join the points c_{2k-1} and c_{2k}, $k = 1, \ldots, \nu$, by means of a curve Q_k; here the ν curves Q_k are supposed to be disjoint. Next we cut two copies of the z-plane along the curves $Q_1, \ldots, Q_\nu$, and join the two sheets crosswise along corresponding edges to produce a two-sheeted Riemann surface with the n branch points $c_1, \ldots, c_n$. If n is odd we must add to the branch points $c_1, \ldots, c_n$, the branch point ∞.

We wish to form an intuitive idea of the connectivity of the Riemann surfaces discussed in this section. We start with $\mathfrak{R}_2$ and think of it as covering a sphere. We separate $\mathfrak{R}_2$ into its two constituent cut sheets with edges L_1 and L_2. Now it is clear that each of these sheets can be continuously deformed into a hemisphere. Let $\mathfrak{B}_1^*$ and $\mathfrak{B}_2^*$ be the two hemispheres in question, and let L_1^* and L_2^* denote the arcs forming the boundaries of these hemispheres and originating in the edges L_1 and L_2 (Figure 12). If these boundary arcs are again connected in the manner of the Riemann surface $\mathfrak{R}_2$, the result is simply a spherical surface. This means that it is possible to map the Riemann surface $\mathfrak{R}_2$ of the function $\sqrt{(z-a)(z-b)}$ in a one-one bicontinuous manner onto the surface of a unit sphere $\mathfrak{K}$. The same is true of $\mathfrak{R}_1$. Now, by the Jordan curve theorem (which is relatively easy to prove for polygons and, more generally, for piecewise smooth curves) a simple closed curve separates a plane or a sphere into two disjoint parts. In

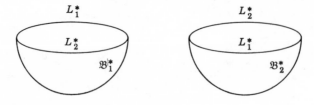

Figure 12

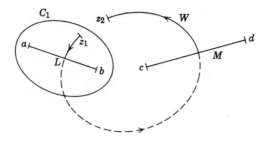

Figure 13

view of the existence of a one-one bicontinuous mapping from $\Re$ to $\Re_2$, it follows that the Riemann surface $\Re_2$ also has the property that each simple closed curve separates it into two parts.

We are about to show that this assertion is false for the Riemann surface $\Re_4$. We take as our simple closed curve a curve C_1 in the z-plane such that the cut L from a to b is interior to C_1 and the cut M from c to d is exterior to C_1. We think of the two sheets $\mathfrak{B}_1$ and $\mathfrak{B}_2$ of $\Re_4$ as lying one above the other and of C_1 as being in the upper sheet $\mathfrak{B}_1$. We shall show that it is possible to connect an interior point z_1 of C_1 to an exterior point z_2 of C_1 by means of a path W which lies on $\Re_4$, is free of double points, and does not intersect the curve C_1 (Figure 13). To this end we go on $\mathfrak{B}_1$ from the point z_1 to a point of the cut L (avoiding C_1), descend to the sheet $\mathfrak{B}_2$, traverse the dotted path indicated in Figure 13 until we reach the cut M (without, of course, intercepting the curve C_1 which lies entirely on the sheet $\mathfrak{B}_1$), and continue in $\mathfrak{B}_1$ from M to z_2 (again avoiding C_1). Thus when the Riemann surface $\Re_4$ is cut along the curve C_1, it is still possible to go from z_1 to z_2.

We shall prove a stronger statement. If z_1 and z_2 approach a point z_0 on C_1, then W goes over into a simple closed curve C_2 on $\Re_4$ which has the single point z_0 in common with C_1 (Figure 14). With $\Re_4$ cut along C_1 and C_2 it is still possible to go from, say, z_3 to z_4 (see Figure 14). This suggests that cutting the surface along C_1 and C_2 leaves it connected. To see this more clearly we effect a transformation of $\Re_4$ which corresponds to the deformation of $\Re_2$ into a sphere $\Re$ (cf. above). We imagine the two sheets $\mathfrak{B}_1$ and $\mathfrak{B}_2$ of $\Re_4$ over the complex z-sphere disconnected along the cuts L and M. These sheets can be continuously deformed into cylinders $\mathfrak{B}_1^*$, $\mathfrak{B}_2^*$. Under such a deformation the edges L_1, L_2 and M_1, M_2 of the cuts L and M go into semicircular arcs L_1^*, L_2^* and M_1^*, M_2^*. If we bend our cylinders into half rings and join corresponding portions of their boundaries we obtain a torus $\mathfrak{T}$ (Figure 15). This shows that the Riemann surface $\Re_4$ can be mapped in a one-one bicontinuous manner onto the surface of a torus so that C_1 goes over into a "longitudinal" circle and C_2 goes over into a "latitudinal"

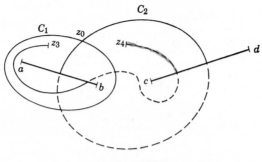

Figure 14

circle on $\mathfrak{T}$. (This last assertion is made plausible by Figure 15.) If we cut $\mathfrak{T}$ along C_1 we can deform it into a cylinder. If we cut the cylinder along C_2 we obtain a connected plane region, namely, a rectangle. This shows, among other things, that when the Riemann surface $\mathfrak{R}_4$ is cut along C_1 and C_2 it remains connected.

A domain and, more generally, a Riemann region $\mathfrak{G}$, is said to be *simply connected* if every closed curve in the region can be contracted to a point of $\mathfrak{G}$. Clearly, a rectangle is simply connected since every closed curve in the rectangle can be contracted to a point by means of affine transformations. It follows that the Riemann surface $\mathfrak{R}_4$ cut along C_1 and C_2 is simply connected.

We managed to map the Riemann surface $\mathfrak{R}_4$ in a one-one bicontinuous manner onto a torus. The same can be done in the case of $\mathfrak{R}_3$. This suggests the following more general problem: Let $\mathfrak{F}$ and $\mathfrak{F}^*$ be two geometric objects in the plane or in space, that is, two sets of points described, say, in

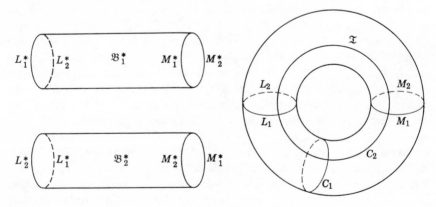

Figure 15

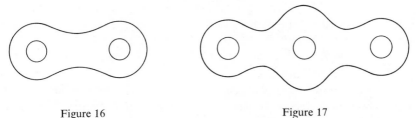

Figure 16   Figure 17

terms of their cartesian coordinates. We are to decide whether it is possible to map one of these objects onto the other in a one-one bicontinuous manner (the continuity requirement means, of course, that a sequence of points $\mathfrak{y}_1, \mathfrak{y}_2, \ldots$ on $\mathfrak{F}$ converges to a point $\mathfrak{y}_0$ on $\mathfrak{F}$ if and only if the sequence $\mathfrak{y}_1^*, \mathfrak{y}_2^*, \ldots$ of the images of $\mathfrak{y}_1, \mathfrak{y}_2, \ldots$ on $\mathfrak{F}^*$ converges to the image $\mathfrak{y}_0^*$ of $\mathfrak{y}_0$). One-to-one bicontinuous mappings are called *topological mappings* or *homeomorphisms*, and topology studies the properties of geometric objects preserved under topological mappings. Two objects which can be mapped topologically onto each other are said to be *homeomorphic*. The problem can be generalized by considering it in the setting of n-dimensional euclidean space or even in more general spaces. We shall not discuss these matters here since we are concerned exclusively with Riemann regions. Incidentally, Riemann was the first to recognize the importance of topology in the theory of functions. The origins of topology go back to Descartes, Leibniz, and Euler.

We saw that it is possible to map $\mathfrak{R}_1$ and $\mathfrak{R}_2$ topologically onto a sphere $\mathfrak{K}$, and $\mathfrak{R}_3$ and $\mathfrak{R}_4$, onto a torus $\mathfrak{T}$. More generally, if $\mathfrak{R}_n$ is the Riemann surface of the function $w = \sqrt{P(z)}$, where $P(z)$ is a polynomial of degree n, $n = 2\nu$ or $n = 2\nu - 1$ with simple zeros, then it is possible to map $\mathfrak{R}_n$ topologically onto a closed surface in space with $\nu - 1$ holes. In particular, if $P(z)$ is a polynomial of degree 5 or 6, then the Riemann surface of $\sqrt{P(z)}$ is homeomorphic to a pretzel (Figure 16), and if $P(z)$ is a polynomial of degree 7 or 8, then the Riemann surface of $\sqrt{P(z)}$ is homeomorphic to a surface with three holes (Figure 17).

6. Elliptic integrals of the first kind

Let a, b, c, d be four distinct complex numbers and let

$$P(z) = (z - a)(z - b)(z - c)(z - d), \qquad Q(z) = \frac{1}{\sqrt{P(z)}}.$$

Let $\mathfrak{R}$ denote the Riemann surface of the function $Q(z)$ and let $\mathfrak{z}$ denote an arbitrary point of $\mathfrak{R}$. If z is a point of the number sphere distinct from

a, b, c, d, then z is covered by two distinct points $\mathfrak{z}$ of $\mathfrak{R}$. The points $z = a, b, c, d$ yield the four branch points of $\mathfrak{R}$. Let $\mathfrak{z}_0$ be a fixed point of $\mathfrak{R}$ lying over a point z_0 and distinct from the branch points as well as the two points at infinity. By associating to $\mathfrak{z}_0$ one of the two distinct values of $1/\sqrt{P(z_0)}$ as the functional value $Q(z_0) = q_0 \neq 0$, we determine a definite branch of $Q(z)$ in the vicinity of $\mathfrak{z}_0$ and so, in the large, a definite function $Q(z)$ on $\mathfrak{R}$. This function is regular and $\neq 0$ at all finite points $\mathfrak{z}$ which are different from the branch points. The expansion

$$(1) \quad Q(t^{-1}) = t^2[(1 - at)(1 - bt)(1 - ct)(1 - dt)]^{-\frac{1}{2}} = \pm t^2 (1 + \cdots)$$

in powers of t shows that the two points at infinity of $\mathfrak{R}$ are double zeros. $Q(z)$ is infinite at the branch points. To determine the order of these poles we must look at the power series expansion in the local parameter. At $z = a$ we put $t = \sqrt{z - a}$. Then

$$(2) \qquad Q(a + t^2) = t^{-1}[(a - b + t^2)(a - c + t^2)(a - d + t^2)]^{-\frac{1}{2}}$$
$$= c_0 t^{-1} + \cdots, \qquad c_0 \neq 0$$

and

$$c_0^{-2} = (a - b)(a - c)(a - d),$$

which shows that the branch points are simple poles.

Let C_1 be a piecewise smooth curve on $\mathfrak{R}$ which joins the fixed point $\mathfrak{z}_0$ to an arbitrary point $\mathfrak{z}_1$ over z_1. We consider the elliptic integral of the first kind

$$(3) \qquad\qquad w = w(C_1) = \int_{C_1} Q(\zeta)\, d\zeta,$$

where ζ varies over the curve which is the projection of C_1 to the number sphere. We must ascertain, in the first place, that this integral has meaning for all curves C_1. It is clear that for this purpose it suffices to investigate the integrand at infinity and at the branch points. Now

$$(4) \qquad\qquad \frac{d(t^{-1})}{dt} = -t^{-2}, \qquad \frac{d(a + t^2)}{dt} = 2t,$$

and (1) and (2) show that the function $Q(z)\, dz/dt$, viewed as a function of the local parameters $t = z^{-1}$ and $t = \sqrt{z - a}$, is regular at $t = 0$. This implies that the integration can be extended to every point of $\mathfrak{R}$.

The function $Q(z)$ is regular in every disk K_0 with center z_0 which does not include one of the points a, b, c, d. In the neighborhood of $\mathfrak{z}_0$ determined by this disk, we define

$$(5) \qquad\qquad w(z) = \int_{z_0}^{z} Q(\zeta)\, d\zeta, \qquad Q(z_0) = q_0,$$

where the integral is taken along the segment joining z_0 to z. If for $Q(\zeta)$ in (5) we put the appropriate series in powers of $\zeta - z_0$ and integrate termwise, then we obtain at $\mathfrak{z}_0$ a function element $w(z)$. We claim that this function element can be continued analytically on $\mathfrak{R}$ along every path C_1, and that the value w of this continuation at the end point $\mathfrak{z}_1$ is given by the integral (3). We note that, in view of Cauchy's theorem, $w = w(z_1)$ when C_1 lies completely over K_0.

Now let $\mathfrak{z}_2$ be a point of C_1 with projection z_2 and let C_2 be the part of C_1 joining $\mathfrak{z}_0$ to $\mathfrak{z}_2$. The local parameter at $\mathfrak{z}_2$ is $t = z - z_2$ for $z_2 \neq a, b, c, d, \infty$, $t = \sqrt{z - z_2}$ for $z_2 = a, b, c, d$, and $t = z^{-1}$ for $z_2 = \infty$. We let $\mathfrak{z}$ vary on C_1 in the vicinity of $\mathfrak{z}_2$ and denote by C the arc of C_1 joining $\mathfrak{z}_0$ to $\mathfrak{z}$. To compute $w(C)$ we integrate along C_1 from $\mathfrak{z}_0$ to $\mathfrak{z}_2$ and then from $\mathfrak{z}_2$ to $\mathfrak{z}$; here we do not exclude the possibility that $\mathfrak{z}$ may lie between $\mathfrak{z}_0$ and $\mathfrak{z}_2$. If $\mathfrak{z}$ is in a sufficiently small neighborhood of $\mathfrak{z}_2$, then $Q(z)\, dz/dt$, viewed as a function of t, is regular there and, by Cauchy's theorem,

$$(6) \qquad w(C) = w(C_2) + \int_0^t Q(z) \frac{dz}{dt}\, dt,$$

where the path in the integral on the right is rectilinear. It follows that the values of w on C_1 close to an arbitrary point $\mathfrak{z}_2$ are given by a regular function of t, and this implies that $w(C)$ coincides in K_0 with $w(z)$.

The equality (6) shows that in order to obtain the function element resulting from the continuation of $w(z)$ along C_1, we must integrate the power series for $Q(z)\, dz/dt$ with respect to the local parameter t. The relations (1), (2), (4) show that at $t = 0$ the derivative

$$\frac{dw}{dt} = Q(z) \frac{dz}{dt} \neq 0.$$

This implies that for every point $\mathfrak{z}_1$ of $\mathfrak{R}$ the mapping effected by the function element w, viewed as a function of the local parameter, is conformal.

7. The inverse function

We proved in the previous section that the function $w(z)$ given locally at $\mathfrak{z}_0$ by (5) can be continued analytically along every curve C on the Riemann surface $\mathfrak{R}$, and the continuation is given by means of the integral

$$(1) \qquad w = w(C) = \int_C Q(\zeta)\, d\zeta.$$

Since the derivative

$$\frac{dw(z)}{dt} = Q(z) \frac{dz}{dt}$$

is different from zero at z_0 and $w(z_0) = 0$, the inverse function $z = f(w)$ of the complex variable w exists and is regular in the vicinity of $w = 0$ and $f(0) = z_0$. We shall see later that the inverse function element $f(w)$ at $w = 0$ can be uniquely continued throughout the w-plane and yields a meromorphic function. To prove this assertion we must so generalize the process of analytic continuation that it is not obstructed by a pole. Suppose the function $f(w)$ is to be continued along a curve C' in the w-plane issuing from 0. Further, suppose that this function is regular on the subarc of C' from 0 to a point w_0 with the exception of w_0, where it has a pole with respect to the local parameter $w - w_0$. Then $f(w)$ is $\neq 0$ for all points w_1 of the subarc which are sufficiently close to w_0. But then the reciprocal function

$$g(w) = \frac{1}{f(w)}$$

is regular on the subarc connecting some w_1 and w_0 including the point w_0 where it has a zero and so can be continued analytically along a portion of C' including w_0. Beyond w_0 we set

$$f(w) = \frac{1}{g(w)}.$$

It would, of course, have been possible to obtain the analytic continuation of $f(w)$ along C' past w_0 by making a slight detour at w_0 to one or to the other side of C' and so avoiding the pole. In this way the process of analytic continuation can be carried out whenever our function has finitely many poles on the curve C' none of which is a branch point.

Now suppose the function $f(w)$ has been continued analytically in the w-plane along a curve C_1' from 0 to its end point w_1. If w_1 is a zero or a pole of f, then we denote by $r(C_1')$ the radius of convergence of the expansion of, respectively, f or $1/f$ in powers of $w - w_1$; otherwise, we denote by $r(C_1')$ the maximum of the two radii of convergence of f and $1/f$. In order to establish the asserted meromorphic character of $f(w)$ we prove, first of all, that if a, b, c, d are held fixed then $r(C_1')$ depends solely on the value $z_1 = f(w_1)$ at the end point of C_1'.

Let C' denote the arc joining the initial point 0 to a variable point w of C_1'. Further, let C and C_1 denote the image curves on the z-sphere of C' and C_1' under the mapping $z = f(w)$, and let z and z_1 be the end points of C and C_1. Since we admit poles of $f(w)$ on C_1', it follows that the point ∞ may lie on C_1. Now, in the vicinity of $w = 0$, $f(w)$ is the inverse of the function $w(z)$ defined by (5) in Section 6 and, therefore,

$$\frac{df(w)}{dw} = \frac{dz}{dw} = \left(\frac{dw}{dz}\right)^{-1} = \frac{1}{Q(z)} = \sqrt{P(z)}.$$

This means that the inverse function $z = f(w)$ satisfies the differential equation

$$(2) \qquad \frac{dz}{dw} = \sqrt{P(z)}$$

with the initial condition $z = z_0$ for $w = 0$; here the branch of the square root is determined by the condition

$$\sqrt{P(z_0)} = \frac{1}{Q(z_0)} = q_0^{-1}.$$

To begin with, this differential equation holds in a sufficiently small neighborhood of $w = 0$, but analytic continuation extends its validity to all of C_1'. This shows that, apart from poles, the derivative $\sqrt{P(z)}$ of the function $f(w)$ is regular on C_1' so that, in particular, this square root is uniquely determined on C_1'. We may now view C_1 and the subarc C as well-defined curves on the Riemann surface $\mathfrak{R}$ issuing from the point $\mathfrak{z}_0$ over z_0. The inverse w of the function $z = f(w)$ is defined on C_1 and, in view of (2), we have on C_1 the relations

$$\frac{dw}{dz} = Q(z), \qquad \frac{dw}{dt} = Q(z) \frac{dz}{dt}.$$

In other words, w is again seen to be the elliptic integral $w = w(C)$ defined by (1). Conversely, in view of the conformal nature of the mapping, for an arbitrary choice of C_1 on $\mathfrak{R}$, C_1' is uniquely determined in the w-plane.

Now, z_1 may be a branch point, or ∞, or it may be different from a, b, c, d, ∞. In the first case let $z_1 = a$, say, so that $t = \sqrt{z - a}$ is the local parameter on $\mathfrak{R}$. In view of (2) and (4) in Section 6, we have

$$Q(z) \frac{dz}{dt} = 2c_0(1 + c_1 t^2 + c_2 t^4 + \cdots), \qquad c_0^{-2} = (a - b)(a - c)(a - d),$$

where the coefficients $c_1, c_2, \ldots$ and the radius of convergence of the power series depend solely on a, b, c, d. Integration yields

$$w - w_1 = 2c_0 t \left(1 + \frac{c_1}{3} t^2 + \frac{c_2}{5} t^4 + \cdots \right),$$

$$\frac{(a - b)(a - c)(a - d)}{4} (w - w_1)^2 = t^2 + d_1 t^4 + d_2 t^6 + \cdots$$

$$= (z - a) + d_1(z - a)^2 + d_2(z - a)^3 + \cdots,$$

and from this relation we obtain by series inversion the function $z - z_1$ as a power series in $(w - w_1)^2$ whose coefficients and radius of convergence depend solely on a, b, c, d. If, in addition, $z_1 \neq 0$, then the same is true of the series expansion of z^{-1} about $w = w_1$. In the second case the local parameter is $t = z^{-1}$. In view of (1) and (4) in Section 6, we now have

$$Q(z) \frac{dz}{dt} = \mp(1 + e_1 t + e_2 t^2 + \cdots),$$

where the coefficients $e_1, e_2, \ldots$ again depend only on a, b, c, d and, furthermore

$$\mp(w - w_1) = t + \frac{e_1}{2} t^2 + \frac{e_2}{3} t^3 + \cdots.$$

Hence in this case the function $t = z^{-1}$ is seen to be a power series in $\mp(w - w_1)$ whose coefficients and radius of convergence depend on a, b, c, d alone; here the two signs correspond to the two points at infinity of $\Re$. In the third case $t = z - z_1$ and

$$w - w_1 = \int_0^t Q(z_1 + t) \, dt, \qquad Q(z_1 + t) = Q(z_1) H(t), \qquad Q(z_1) \neq 0,$$

where

$$H(t) = \left[\left(1 + \frac{t}{z_1 - a}\right) \left(1 + \frac{t}{z_1 - b}\right) \left(1 + \frac{t}{z_1 - c}\right) \left(1 + \frac{t}{z_1 - d}\right) \right]^{-\frac{1}{2}} = 1 + \cdots.$$

Since the coefficients of the series $H(t)$ again depend only on a, b, c, d and z_1, it follows, just as in the previous two cases, that the series expansion of $z - z_1$ and of z^{-1} for $z_1 \neq 0$ in powers of $w - w_1$ has a radius of convergence which depends only on a, b, c, d and z_1. This proves, among other things, our earlier assertion concerning $r(C_1')$ so that from now on we can write

$$r(C_1') = r(z_1).$$

We shall now show that the radius $r(z_1)$ exceeds a positive bound which is independent of z_1. The proof is by contradiction. Specifically, let C_n' ($n = 1, 2, \ldots$) be a sequence of curves in the w-plane issuing from 0 such that the function $z = f(w)$ can be continued analytically on each of these curves and such that

(3) $\lim_{n \to \infty} r(C_n') = 0.$

The image curves C_n are uniquely determined on $\Re$ and join $\mathfrak{z}_0$ to $\mathfrak{z}_n$. Since $\Re$ is compact we may assume (after possibly shifting to a subsequence) that the points $\mathfrak{z}_n$ ($n = 1, 2, \ldots$) converge to a point $\mathfrak{z}_\infty$ on $\Re$. Let z_n and z_∞ be the projections of $\mathfrak{z}_n$ and $\mathfrak{z}_\infty$. Let S_n denote the polygonal curve on $\Re$ joining $\mathfrak{z}_0$ to $\mathfrak{z}_1$ to $\ldots$ to $\mathfrak{z}_{n-1}$ to $\mathfrak{z}_n$. Let $\mathfrak{U}$ be a disk about $\mathfrak{z}_\infty$ so small that $Q(z) \, (dz/dt)$ is regular throughout $\mathfrak{U}$ when viewed as a function of the local parameter at $\mathfrak{z}_\infty$. Next choose m so large that all the points $\mathfrak{z}_n$ with $n > m$

lie in $\mathfrak{U}$. Adjoin to S_n the segment with end points $\mathfrak{z}_n$, $\mathfrak{z}_\infty$ and denote by T_n the resulting polygonal curve joining $\mathfrak{z}_0$ to $\mathfrak{z}_\infty$. In view of Cauchy's theorem the value of the integral

$$w_\infty = \int_{T_n} Q(\zeta)\, d\zeta \qquad (n > m)$$

is independent of n. If we put

$$w_n = \int_{S_n} Q(\zeta)\, d\zeta \qquad (n = 1, 2, \ldots),$$

it follows that

$$\lim_{n \to \infty} w_n = w_\infty.$$

Let T_n' be the image in the w-plane of the curve T_n under the elliptic integral w. Continue the inverse function $z = f(w)$ along T_n' to $w = w_\infty$. If $z_\infty \neq 0$, ∞, then we obtain for z and z^{-1} two power series expansions whose radii of convergence have the maximal value

(4) $$r(z_\infty) = r > 0.$$

If $z_\infty = 0$ or $z_\infty = \infty$, then we have, respectively, for z or z^{-1} a series with radius of convergence r. Also, the number r defined in (4) is independent of n. For n sufficiently large we have on the arc of T_n' between w_n and w_∞ the estimate

$$|w - w_\infty| < \frac{r}{2}.$$

In view of the definition of the radius r, the function element f or f^{-1} obtained at w_n as a result of analytic continuation along T_n' is regular in the disk

$$|w - w_n| < \frac{r}{2}$$

(Figure 18). But then

$$r(C_n') = r(z_n) = r(T_n') \geq \frac{r}{2},$$

which for $n \to \infty$ contradicts the assumption (3).

It is now easy to show that it is possible to continue analytically the inverse function element $z = f(w)$ from 0 along any curve C' in the w-plane. To see this we need only bear in mind the fact that the radius of convergence $r(z_1)$ exceeds some positive bound ρ independent of z_1, so that, having arrived at a point w_1 of C' in the process of analytic continuation from 0 along C', we can continue into a circle of radius ρ and center w_1 and can in this way reach the end point of C' in a finite number of steps. Here it was essential to employ

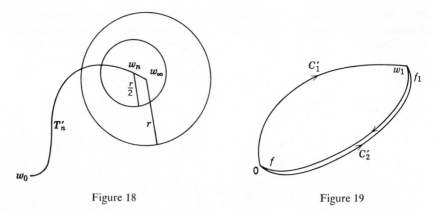

Figure 18   Figure 19

the generalized concept of analytic continuation which includes analytic continuation past a pole.

Finally it remains to show that the analytic continuation of the function $f(w)$ from $w = 0$ to an arbitrary point w_1 is independent of the path joining 0 to w_1. First let $w_1 = 0$ and let C' be a closed curve in the w-plane leading from 0 to 0. If the curve C' is continuously deformed while its point 0 is held fixed, then we can assert on the basis of the theorem proved at the end of Section 3 that analytic continuation along the altered curve of the function element $f(w)$ given at $w = 0$ yields the same result as analytic continuation along C'. Since the w-plane is simply connected it is possible to shrink C' continuously to the single point 0, and this shows that analytic continuation to $w_1 = 0$ is independent of the closed path C'. To prove the corresponding assertion for $w_1 \neq 0$, we consider two curves C_1' and C_2' joining 0 to w_1. By going from 0 to w_1 along C_1' and then back to 0 along C_2' in the opposite direction we form a closed curve C' to which we can apply the conclusion just established. If we continue the function element f given at 0 analytically along C', then we obtain at w_1 the power series f_1 which results from continuation along the arc C_1', and after traversing the remaining arc of C' we obtain at $w = 0$ the original function element f. If we now go again on C_1' from 0 to w_1, then, clearly, f goes over into f_1, and this shows that the continuation of f along C_1' and C_2' leads to the same function element f_1 (Figure 19). In just the same way we prove, more generally, the monodromy theorem of function theory which asserts that a function element which can be continued analytically from a point of a simply connected Riemann region $\mathfrak{G}$ along all paths defines a single-valued function on $\mathfrak{G}$.

Since analytic continuation of the function element $z = f(w)$ given at 0 is independent of the path in the w-plane, we may use the notation

$$z = f(w)$$

in the large without specifying the path from 0 to w. It follows that the function $f(w)$ is actually meromorphic, that is, regular in the whole w-plane except for poles. In the following sections we shall investigate $f(w)$ more searchingly. One important property of this function which we already know is that it satisfies the differential equation

$$\left(\frac{dz}{dw}\right)^2 = P(z), \qquad P(z) = (z - a)(z - b)(z - c)(z - d),$$

which expresses the fact that $z = f(w)$ is the result of inverting an elliptic integral of the first kind. In the sequel we shall be primarily concerned with the study of the periodicity properties of the function $f(w)$.

8. The covering surface

At this point it is expedient to introduce the concept of homotopy and to define the so-called fundamental group; a detailed presentation of these ideas is found in books on topology. Two directed curves on a surface $\mathfrak{F}$ are called *homotopic* if they can be continuously deformed into each other on $\mathfrak{F}$ while their beginning points and end points remain fixed; this makes it clear that the beginning point and end point of one curve must coincide with the beginning point and end point of the other curve. We shall not give a precise definition of the concept of surface; the term will refer to a concretely given connected surface in space or to a Riemann region over the z-sphere. If A and B are two homotopic curves then we write $A \sim B$. It is clear that $A \sim B$ implies $B \sim A$, and $A \sim B$ and $B \sim C$ imply $A \sim C$. If the beginning point of a curve A_2 is the end point of a curve A_1 then $A_1 A_2$ denotes the joint curve where we first traverse A_1 and then A_2. If $A_1 \sim B_1$ and $A_2 \sim B_2$ then $B_1 B_2$ has meaning and, as is readily shown, $A_1 A_2 \sim B_1 B_2$; in fact, to deform $A_1 A_2$ into $B_1 B_2$, we keep the beginning point and end point of A_1 fixed and deform A_1 continuously into B_1, and then, similarly, deform A_2 into B_2. It is clear that our multiplication of curves is associative, that is, $(A_1 A_2)A_3 = A_1(A_2 A_3)$, and this justifies the notation $A_1 A_2 A_3$. Here we must, of course, see to it that such a product makes sense, that is, that for two neighboring factors the end point of the left factor coincides with the beginning point of the right factor. Finally A^{-1} denotes the curve obtained from a curve A by reversing its direction. We have

$$(A^{-1})^{-1} = A, \qquad (AB)^{-1} = B^{-1}A^{-1},$$

and $A \sim B$ implies $A^{-1} \sim B^{-1}$.

In particular, we consider on the surface $\mathfrak{F}$ curves $A, B, \ldots$ with common beginning and end point $\mathfrak{y}$, that is, closed curves issuing from $\mathfrak{y}$. Here we include the degenerate curve E consisting of the single point η. The curve E has the property

$$AE = EA = A.$$

If $A \sim E$ then we say that A *is homotopic to zero.* If C is a (not-necessarily-closed) curve on $\mathfrak{F}$ issuing from η, then the curve CC^{-1} is closed, and can be trivially contracted to η so that $CC^{-1} \sim E$. It follows that $C \sim D$ is equivalent to $DC^{-1} \sim E$. We separate the closed curves issuing from η into homotopy classes $\alpha, \beta, \ldots$ where A_1 and A_2 belong to the same class if and only if $A_1 \sim A_2$. It is natural to define the product $\alpha\beta$ of two homotopy classes as the class containing the product AB of a curve A in the class α and a curve B in the class β; the product $\alpha\beta$ is seen to be independent of the particular choice of curves A and B. The previous rules show that the homotopy classes form a group under multiplication with unit element ε consisting of the curves homotopic to zero. The resulting group Γ is called the *fundamental group* of the given surface. The group Γ seems to depend on the given point η. We must see to what extent this is so.

Let η^* be another point of the surface and let H be a fixed curve on $\mathfrak{F}$ joining η to η^*. If A^* is a closed curve issuing from η^* then $HA^*H^{-1} = A$ is a closed curve issuing from η. If $A^* \sim B^*$ and $B = HB^*H^{-1}$, then $A \sim B$. We see that the relation $HA^*H^{-1} = A$ associates with every homotopy class α^* relative to η^* a unique homotopy class α relative to η. Conversely, if A is a closed curve issuing from η, then $A^* = H^{-1}AH$ is a closed curve issuing from η^* and we have

$$HA^*H^{-1} = HH^{-1}AHH^{-1} \sim EAE = A.$$

It follows that our correspondence is one-one. Since, in addition,

$$(A_1A_2)^* = H^{-1}A_1A_2H \sim H^{-1}A_1HH^{-1}A_2H = A_1^*A_2^*$$

we conclude that the correspondence from A to A^* defines an isomorphism of the two fundamental groups. This shows that the homotopy group of a surface is determined up to an isomorphism by the surface alone. In particular we see that a closed curve issuing from η is homotopic to zero if and only if, with η no longer fixed, it is possible to contract the curve continuously on the surface to any point. It follows that a surface is simply connected if and only if its fundamental group consists of the identity alone.

In the sequel we will determine the fundamental group of the Riemann surface $\mathfrak{R} = \mathfrak{R}_4$ in two ways, and now we again consider the elliptic integral

$$w = w(C) = \int_C Q(\zeta)\,d\zeta,$$

where C joins the fixed point $\mathfrak{z}_0$ to the variable point $\mathfrak{z}$ on $\mathfrak{R}$. If C, A, B are closed curves issuing from $\mathfrak{z}_0$ then, clearly,

$$w(C^{-1}) = -w(C), \qquad w(AB) = w(A) + w(B).$$

Theorem 1: If C_1 and C_2 are two curves on $\mathfrak{R}$ issuing from $\mathfrak{z}_0$, then $w(C_1) = w(C_2)$ if and only if $C_1 \sim C_2$.

Proof: Let $\mathfrak{z}_1$ and $\mathfrak{z}_2$ be the end points of C_1 and C_2 and let C denote the subarc of C_1 joining $\mathfrak{z}_0$ to a variable point $\mathfrak{z}$. As $\mathfrak{z}$ traverses the curve C_1, the point $w = w(C)$ traverses an image curve C_1' in the w-plane which extends from 0 to $w_1 = w(C_1)$. Conversely, C_1' determines uniquely the curve C_1 on $\mathfrak{R}$ by means of the relation

$$(1) \qquad\qquad z = f(w), \qquad \frac{df(w)}{dw} = \sqrt{P(z)},$$

and the value w determines the point $\mathfrak{z}$. Similarly, let C_2' denote the image of C_2. If the end points $w_1 = w(C_1)$ and $w_2 = w(C_2)$ of C_1' and C_2' coincide, then, the plane being simply connected, $C_1' \sim C_2'$ in the plane and we have $\mathfrak{z}_1 = \mathfrak{z}_2$. If we deform C_1' to C_2' keeping their common end points fixed, then (1) yields a corresponding continuous deformation of C_1 to C_2. Conversely, if $C_1 \sim C_2$ on $\mathfrak{R}$, then Cauchy's theorem justifies the conclusion that the integral $w(C)$ is not changed when C is continuously deformed on $\mathfrak{R}$ while the beginning point and end point of C are held fixed. This implies that $w(C_1) = w(C_2)$, and the proof of the theorem is complete.

Using Theorem 1 we can readily determine the Riemann surface $\mathfrak{S}$ of the analytic function defined in the neighborhood of $\mathfrak{z}_0$ by the function element $w(z)$. To this end we consider two paths C_1 and C_2 issuing from $\mathfrak{z}_0$ and leading to the same point $\mathfrak{z}_1$. Theorem 1 states that analytic continuation of the function element $w(z)$ at $\mathfrak{z}_0$ yields the same value $w(C_1) = w(C_2)$ at $\mathfrak{z}_1$ if and only if $C_1 \sim C_2$. If we vary $\mathfrak{z}_1$ in the vicinity of a given point $\mathfrak{z}$, then it follows that the analytic continuation of $w(z)$ along two curves C_1 and C_2 on $\mathfrak{R}$ which lead from $\mathfrak{z}_0$ to $\mathfrak{z}$ yields the same function element at $\mathfrak{z}$ if and only if $C_1 \sim C_2$. This means that the analytic continuation depends in a one-to-one manner on the homotopy class of the path. We now select in each homotopy class α of closed paths a representative C_α and form the paths $C_\alpha C_1$, C_1 fixed. The paths $C_\alpha C_1$ yield a complete system of nonhomotopic paths from $\mathfrak{z}_0$ to $\mathfrak{z}_1$. These paths are in one-one correspondence with the totality of different analytic continuations of the function w from $\mathfrak{z}_0$ to $\mathfrak{z}_1$, and for each class α we obtain over the arbitrarily selected point $\mathfrak{z}_1$ of $\mathfrak{R}$ a surface element of the required Riemann surface $\mathfrak{S}$. In the sequel we shall investigate the construction of $\mathfrak{S}$ more closely. We note that in view of Theorem 1 the function $w(z)$ is not only single-valued on $\mathfrak{S}$ but also invariably takes on different values at different points of $\mathfrak{S}$. Since the variable w in the inverse function $z = f(w)$ can vary over the whole complex plane, it follows that w maps the surface $\mathfrak{S}$ conformally onto the plane. We call $\mathfrak{S}$ the *covering surface* of $\mathfrak{R}$, and can give our findings the following form:

Theorem 2: The elliptic integral w maps the covering surface of the Riemann surface $\mathfrak{R}$ in a one-to-one conformal manner onto the full plane.

Since the plane is simply connected, Theorem 2 permits us to draw the same conclusion about the covering surface $\mathfrak{S}$ of the Riemann surface $\mathfrak{R}$.

9. The periods

If A is a closed curve issuing from a point $\mathfrak{z}_0$ on $\mathfrak{R}$, then we call the integral

$$(1) \qquad\qquad w(A) = \int_A Q(\zeta)\, d\zeta$$

the *period* along A. In particular $w(E) = 0$ for the degenerate curve E consisting of the single point $\mathfrak{z}_0$. In view of Theorem 1 in the previous section, the periods $w(A)$ and $w(B)$ along two closed curves A and B issuing from $\mathfrak{z}_0$ are equal if and only if A and B belong to the same homotopy class. In particular, $w(A) = 0$ if and only if A is homotopic to zero. It follows that with each homotopy class α there is associated a period $w(\alpha)$ and this correspondence is one-one; also $w(\varepsilon) = 0$. Since

$$w(\alpha\beta) = w(\alpha) + w(\beta) = w(\beta) + w(\alpha) = w(\beta\alpha),$$

we conclude that

$$\alpha\beta = \beta\alpha,$$

that is, the fundamental group Γ of $\mathfrak{R}$ is abelian, and the periods $\omega = w(\alpha)$ yield a faithful representation of Γ as an additive vector group. If C is a curve on $\mathfrak{R}$ which joins $\mathfrak{z}_0$ to an arbitrary point $\mathfrak{z}$, then the integral in (1) taken over $A^* = C^{-1}AC$ yields the value $w(A)$, and this shows that the representation of Γ by means of the periods is independent of the initial point $\mathfrak{z}_0$.

We know already that the Riemann surface $\mathfrak{R}$ is not simply connected, but we can now give an alternative function-theoretical proof of this fact. Suppose $\mathfrak{R}$ were simply connected. Then the fundamental group Γ would consist of the identity element ε alone, and the value of the elliptic integral $w(C)$ would depend solely on the end point $\mathfrak{z}$ of C. But then the function element $w(z)$ at $\mathfrak{z}_0$ would admit unique analytic continuation to all points of $\mathfrak{R}$ and w would be single-valued and everywhere regular on $\mathfrak{R}$. Since $\mathfrak{R}$ is compact, the absolute value of w would take on a maximal value at some point of $\mathfrak{R}$. But this is impossible because w is not constant. This contradiction proves that $\mathfrak{R}$ is not simply connected and that not all periods $w(A)$ have the value 0.

Let A be fixed and let C and the end point $\mathfrak{z}$ be variable. The path $AC = D$ again joins $\mathfrak{z}_0$ to $\mathfrak{z}$. If we put

$$w(A) = \omega, \qquad w(C) = w,$$

then

$$w(D) = w + \omega,$$

and we have for the inverse function $z = f(w)$ the relation

$$(2) \qquad f(w) = f(w + \omega).$$

Here $\mathfrak{z}$, and therefore w, can vary, and we conclude that ω is a period in the usual sense of the word of the meromorphic inverse function $f(w)$. Since ω depends only on the class of A, we can define $w(\alpha) = \omega$. On the other hand, if (2) holds for an appropriate constant ω and variable w, then the derivative of $f(w)$ has the same values at w and at $w + \omega$ and there exist two curves C and D on $\mathfrak{R}$ leading from $\mathfrak{z}_0$ to the same end point $\mathfrak{z}$ such that

$$w = w(C), \qquad w + \omega = w(D).$$

For the curve $A = DC^{-1}$ issuing from $\mathfrak{z}_0$ we have $w(A) = w(D) - w(C) = \omega$. This shows that $f(w)$ has no periods other than those already specified.

The periods of a nonconstant meromorphic function cannot have a finite limit point, for in every neighborhood of such a point the function would take on the same value infinitely often, and this is impossible. It follows that the set of nonzero periods must contain a period ω_1 of minimal absolute value. Now let ω be a period for which the quotient ω/ω_1 is real, and let m be an integer such that

$$0 \le \frac{\omega}{\omega_1} - m < 1.$$

Then the difference

$$\omega - m\omega_1 = \omega_0$$

is also a period and $|\omega_0| < |\omega_1|$. This implies that $\omega_0 = 0$, $\omega = m\omega_1$. In other words, the ω for which the quotient ω/ω_1 is real are precisely the multiples $m\omega_1$ $(m = 0, \pm 1, \ldots)$ of ω_1. We claim that these values do not account for all the periods of the function $f(w)$. Suppose this were not the case, and consider the function

$$(3) \qquad v = e^{(2\pi i/\omega_1)w}.$$

Let C and D be two paths on $\mathfrak{R}$ from $\mathfrak{z}_0$ to the same end point $\mathfrak{z}$. Then $DC^{-1} = A$ is closed and

$$(4) \qquad w(D) = w(C) + w(A),$$

where the period $w(A)$ would be a multiple of ω_1. In view of (3) and (4), v would be a single-valued and regular function on $\Re$ and this would result in the same contradiction as above. We conclude that there must exist further periods of $f(w)$ whose ratio to ω_1 is not real. Let ω_2 be a period of this type of minimal absolute value. Since we can replace ω_2 by $-\omega_2$, we may assume that the imaginary part of the quotient

$$\tau = \frac{\omega_2}{\omega_1}$$

is positive; also

$$|\tau| \geq 1.$$

We choose ω_1 and ω_2 as the units of an oblique coordinate system. Then every complex number has a unique representation $x\omega_1 + y\omega_2$ and, furthermore, $x = \xi + \rho$, $y = \eta + \sigma$ with ξ and η integers and

$$-\tfrac{1}{2} \leq \rho < \tfrac{1}{2}, \qquad -\tfrac{1}{2} \leq \sigma < \tfrac{1}{2}.$$

If $\omega = x\omega_1 + y\omega_2$ is a period, then

$$\omega - \xi\omega_1 - \eta\omega_2 = \rho\omega_1 + \sigma\omega_2 = \omega_0$$

is also a period and

$$\left| \frac{\omega_0}{\omega_2} \right| = |\rho\tau^{-1} + \sigma| \leq |\rho|\,|\tau|^{-1} + |\sigma| \leq \tfrac{1}{2} + \tfrac{1}{2} = 1.$$

Here we have in at least one place the inequality $<$, since $\rho\tau^{-1}$ is real only for $\rho = 0$. It follows that the sharper inequality

$$|\omega_0| < |\omega_2|$$

holds, and this means that $\omega_0\omega_1^{-1}$ is real, $\sigma = 0$, $\omega_0\omega_1^{-1} = \rho = 0$, $\omega_0 = 0$, so that

$$\omega = \xi\omega_1 + \eta\omega_2$$

with integers ξ, η. We have therefore proved

Theorem 1: The periods of the elliptic integral $w(C)$ form a lattice spanned by the two basic periods ω_1 and ω_2.

This result justifies calling the inverse function $z = f(w)$ a *doubly periodic function*.

Let α_1 and α_2 be the homotopy classes for which

$$w(\alpha_1) = \omega_1, \qquad w(\alpha_2) = \omega_2.$$

Since the additive group of periods is isomorphic to the fundamental group Γ, each element of Γ can be uniquely represented as a product $\alpha_1^k\alpha_2^l$, k, l integers. This means that α_1 and α_2 are the two commuting and independent generators of Γ. We shall now derive this important result geometrically. Since the concept of homotopy involves only continuous deformation and $\mathfrak{R}$ can be mapped in a one-one bicontinuous manner onto the torus $\mathfrak{T}$, it suffices to construct the fundamental group for the more transparent surface $\mathfrak{T}$. Our aim, then, is to find a complete system of closed curves issuing from a fixed point, no two of which are homotopic; these curves will represent the various homotopy classes.

We have already cut the torus $\mathfrak{T}$ along the longitudinal circle C_1 and the latitudinal circle C_2 and obtained a rectangle with opposite sides identified in pairs. We shall not deal with this identification but rather consider the plane lattice based on this rectangle as mesh unit. To simplify matters we assume that the rectangle has been mapped by means of an affine transformation onto a unit square which we choose as the unit square of a (rectangular) cartesian coordinate system. Next the plane is covered by squares $\mathfrak{Q}_{mn}$ determined by

$$m \leq x \leq m + 1, \qquad n \leq y \leq n + 1$$

with m and n varying independently over the integers. Each square $\mathfrak{Q}_{mn}$ with opposite sides identified is a one-one bicontinuous image of the torus $\mathfrak{T}$. Let $\mathfrak{z}_0$ on $\mathfrak{R}$ be the common point of the cuts C_1 and C_2. The image of $\mathfrak{z}_0$ is some lattice point of the (x, y)-plane, say, the origin $\mathfrak{o}$. Now let A be a closed curve on $\mathfrak{R}$ issuing from $\mathfrak{z}_0$ and let us follow its plane image curve A' issuing from $\mathfrak{o}$. The curve enters one of the four squares at $\mathfrak{o}$, say, the square $\mathfrak{Q}$. We follow it in $\mathfrak{Q}$ until it reaches one of the four sides of this square. In view of the identification rule we would then, strictly speaking, be obliged to shift to the opposite side of $\mathfrak{Q}$ and from there again go into the interior of $\mathfrak{Q}$ (Figure 20). Instead, we extend the image curve into the appropriate adjoining square which, we recall, is also an image of $\mathfrak{T}$. By continuing this procedure we obtain a plane image curve A' of A which issues from $\mathfrak{o} = (0, 0)$ and, since A is closed, terminates at a lattice point $\mathfrak{y} = (m, n)$ (Figure 21); note that $\mathfrak{y}$ need not be the origin. Since the plane is simply connected we can, keeping the initial point $\mathfrak{o}$ and the end point $\mathfrak{y}$ fixed, deform A' continuously into the curve made up of the segments from $(0, 0)$ to $(m, 0)$ and from $(m, 0)$ to (m, n) and is, clearly, the image of the curve $C_1^m C_2^n$ on $\mathfrak{R}$. On the other hand, this continuous deformation of A' can be shifted to the Riemann surface $\mathfrak{R}$. In this way we obtain the continuous deformation $C_1^m C_2^n$ of A. It is clear that, conversely, the image of every continuous deformation of A must terminate at $\mathfrak{y}$. This shows that the curves $C_1^m C_2^n$, with m and n varying

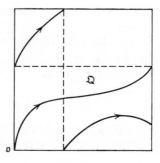

Figure 20

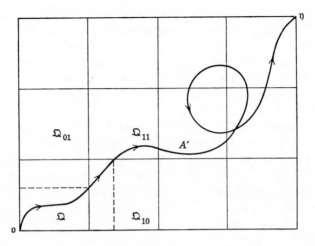

Figure 21

independently over the integers, are precisely the required complete system of representatives of the homotopy classes. Since the images of C_1C_2 and C_2C_1 terminate at $(1, 1)$, it follows that $C_1C_2 \sim C_2C_1$ on $\Re$. If β_1 and β_2 denote the classes of C_1 and C_2, then the fundamental group Γ is precisely the abelian group with free generators β_1 and β_2. If we put $w(\beta_1) = w_1$ and $w(\beta_2) = w_2$, then w_1 and w_2 are both $\neq 0$ and, furthermore,

$$w(\beta_1^m \beta_2^n) = mw_1 + nw_2.$$

The ratio of the periods w_1 and w_2 is not real; otherwise there would exist a period $\omega \neq 0$ of minimal absolute value of which $w_1 = k\omega$ and $w_2 = l\omega$ would be multiples. But then we would have

$$lw_1 = kw_2, \qquad w(\beta_1^l) = w(\beta_2^k), \qquad \beta_1^l = \beta_2^k,$$

with $kl \neq 0$, which is a contradiction. This yields a new proof of our earlier findings.

It remains to clarify the connection between bases α_1, α_2 and β_1, β_2 of the fundamental group Γ of $\Re$. To begin with, let β_1, β_2 denote two arbitrary elements of Γ. Then

(5) $\beta_1 = \alpha_1^m \alpha_2^n, \qquad \beta_2 = \alpha_1^p \alpha_2^q,$

with uniquely determined integer exponents m, n, p, q. We wish to show that β_1, β_2 are a basis of Γ if and only if

(6) $mq - np = \pm 1.$

It is clear that β_1 and β_2 are a basis of Γ if and only if there exist integers m', n', p', q' such that

(7) $$\alpha_1 = \beta_1^{m'}\beta_2^{n'}, \qquad \alpha_2 = \beta_1^{p'}\beta_2^{q'}.$$

In view of (5), (7) is equivalent to

$$\alpha_1 = \alpha_1^{mm'+pn'}\alpha_2^{nm'+qn'}, \qquad \alpha_2 = \alpha_1^{mp'+pq'}\alpha_2^{np'+qq'},$$

that is, bearing in mind the independence of α_1 and α_2, to

$$1 = mm' + pn', \qquad 0 = nm' + qn', \qquad 0 = mp' + pq', \qquad 1 = np' + qq'.$$

These equations can be given the matrix form

(8) $$\begin{pmatrix} 1 & 0 \\ 0 & 1 \end{pmatrix} = \begin{pmatrix} m' & n' \\ p' & q' \end{pmatrix}\begin{pmatrix} m & n \\ p & q \end{pmatrix}.$$

The determinants of the matrices in (8) must satisfy

(9) $$1 = (m'q' - n'p')(mq - np).$$

Since all the numbers in (9) are integers, it follows that each of the numbers in brackets must have the value ± 1, which is the assertion in (6). On the other hand, if (6) holds then (8) yields the relation

$$\begin{pmatrix} m' & n' \\ p' & q' \end{pmatrix} = \begin{pmatrix} m & n \\ p & q \end{pmatrix}^{-1} = \pm\begin{pmatrix} q & -n \\ -p & m \end{pmatrix},$$

and thus the required integer solution of (7).

Since the additive group of periods is isomorphic to the fundamental group, the preceding argument also shows how to obtain from the given basic periods ω_1, ω_2 every pair of basic periods ω_1^*, ω_2^*. Call a homogeneous linear substitution *unimodular* if its coefficients are integers and its determinant is ± 1. We have proved

Theorem 2: Two complex numbers ω_1^*, ω_2^* are basic periods of the function $f(w)$ if and only if they are related to the basic periods ω_1, ω_2 by a unimodular substitution

$$\omega_1^* = m\omega_1 + n\omega_2, \qquad \omega_2^* = p\omega_1 + q\omega_2.$$

It is, of course, possible to prove this result directly without involving the fundamental group.

Earlier we agreed to choose ω_2 so that the ratio $\tau = \omega_2/\omega_1$ has positive imaginary part. Now we ask under what additional condition this holds for $\tau^* = \omega_2^*/\omega_1^*$. With horizontal bars denoting complex conjugation we have

$$\begin{pmatrix} \overline{\omega_1^*} & \omega_1^* \\ \overline{\omega_2^*} & \omega_2^* \end{pmatrix} = \begin{pmatrix} m & n \\ p & q \end{pmatrix}\begin{pmatrix} \overline{\omega}_1 & \omega_1 \\ \overline{\omega}_2 & \omega_2 \end{pmatrix},$$

so that

$$w_1^* \overline{\omega_1^*}(\tau^* - \overline{\tau^*}) = \overline{\omega_1^*}\omega_2^* - \overline{\omega_2^*}\omega_1^* = (mq - np)(\bar{\omega}_1\omega_2 - \bar{\omega}_2\omega_1)$$
$$= (mq - np)\omega_1\bar{\omega}_1(\tau - \bar{\tau}).$$

It follows that the quotient ω_2^*/ω_1^* of the basic periods ω_1^*, ω_2^* has positive imaginary part if and only if the substitution is properly unimodular, that is, has determinant 1.

Originally we obtained a pair of basic periods ω_1, ω_2 by requiring that ω_1 is a nonzero period of minimal absolute value and that, with the imaginary part of the ratio $\tau = \omega_2/\omega_1$ positive, the absolute value of ω_2 is as small as possible. Such a pair of periods is called *reduced*. At first it would seem that determination of a pair of reduced periods might require infinitely many conditions since there are infinitely many periods ω to be tested. Actually, as we are about to show, we manage with three necessary and sufficient conditions. Let ξ and η be the real and imaginary parts of $\tau = \xi + i\eta$, so that $\eta > 0$. We have

Theorem 3: A pair of basic periods ω_1, ω_2 is reduced if and only if the point $\tau = \omega_2/\omega_1 = \xi + i\eta$ lies in the region of the upper half of the complex plane defined by the three inequalities

(10) $$\xi^2 + \eta^2 \geq 1, \qquad -\tfrac{1}{2} \leq \xi \leq \tfrac{1}{2}.$$

Proof: Suppose the pair ω_1, ω_2 reduced. Then, in particular,

$$|\omega_2 \pm \omega_1| \geq |\omega_2| \geq |\omega_1|,$$
$$(\xi \pm 1)^2 + \eta^2 = |\tau \pm 1|^2 \geq |\tau|^2 = \xi^2 + \eta^2 \geq 1;$$

hence, also

$$\pm 2\xi + 1 \geq 0,$$

which implies the three asserted inequalities.

We now show that, conversely, these inequalities imply that the pair ω_1, ω_2 is reduced. Let $\omega = m\omega_1 + n\omega_2$ be a nonzero period. If $n = 0$, then the quotient ω/ω_1 is real and $|\omega| \geq |\omega_1|$. Now let $n \neq 0$. We have

$$\Delta = \left|\frac{\omega}{\omega_1}\right|^2 - \left|\frac{\omega_2}{\omega_1}\right|^2 = |m + n\tau|^2 - |\tau|^2 = (m + n\xi)^2 - \xi^2 + (n^2 - 1)\eta^2.$$

For $n \neq \pm 1$, we have $n^2 - 1 \geq 3$, and

$$\eta^2 = (\xi^2 + \eta^2) - \xi^2 \geq 1 - \tfrac{1}{4} = \tfrac{3}{4}$$

implies that

$$\Delta \geq -\tfrac{1}{4} + \tfrac{9}{4} = 2 > 0.$$

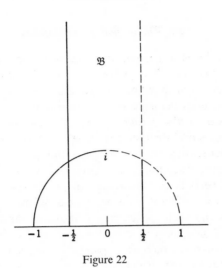

Figure 22

At any rate, for $n = \pm 1$ and for m an arbitrary integer we have

$$\Delta = (m \pm \xi)^2 - \xi^2 \geq 0.$$

This shows that if the quotient ω/ω_1 is not real then, in fact, $|\omega| \geq |\omega_2|$.

Now we propose to sharpen Theorem 3 somewhat. The boundary of the region $\mathfrak{B}$ defined by the inequalities (10) consists of the ray $\xi = -\frac{1}{2}$, $\eta \geq \frac{1}{2}\sqrt{3}$, the ray $\xi = \frac{1}{2}$, $\eta \geq \frac{1}{2}\sqrt{3}$, and the arc $\xi^2 + \eta^2 = 1$, $-\frac{1}{2} \leq \xi \leq \frac{1}{2}$ of the unit circle. If the point τ lies on the second ray, then $\tau - 1$ lies on the first ray. If τ lies on the part of the circular arc with $\xi > 0$, then $-\tau^{-1}$ lies on the part with $\xi < 0$. Since the transition from $\tau = \omega_2/\omega_1$ to $\tau - 1 = (\omega_2 - \omega_1)/\omega_1$ and $-\tau^{-1} = -\omega_1/\omega_2$ is effected by the unimodular substitutions $\omega_1^* = \omega_1^*$, $\omega_2^* = -\omega_1 + \omega_2$ and $\omega_1^* = \omega_2$, $\omega_2^* = -\omega_1$, there always exists a pair of basic periods when we require in (10) that $\xi < \frac{1}{2}$ and that, for $\xi^2 + \eta^2 = 1$, $\xi \leq 0$ (Figure 22). Now we prove that these conditions determine τ uniquely. Indeed, if $\omega_1^* = m\omega_1 + n\omega_2$ and $\omega_2^* = p\omega_1 + q\omega_2$ are such basic periods and $\tau^* = \omega_2^*/\omega_1^*$ denotes their ratio, then $|\omega_1^*| = |\omega_1|$ and $mq - np = 1$. $n = 0$ implies $m = q = \pm 1$, $\omega_1 = \pm \omega_1^*$, $\tau^* = \pm p + \tau$, so that $p = 0$ and $\tau = \tau^*$. $n \neq 0$ implies first, as in the proof of Theorem 3, that $n = \pm 1$ and then $|\omega_1^*| = |\omega_2| = |\omega_1|$, $|\tau| = 1$, $\xi \leq 0$, $m = 0$ or ± 1. If $m = 0$, then $p = \mp 1$, $\omega_1^* = \pm \omega_2$, $\omega_2^* = \mp \omega_1 + q\omega_2$, $\tau^* = -\tau^{-1} \pm q$ and either $q = 0$, $\tau^* = \tau = i$ or $q = \mp 1$, $\xi = -\frac{1}{2}$, $\tau^* = \tau = e^{2\pi i/3}$. On the other hand, if $m = \pm 1$, then we have $q - p = \pm 1$, $\xi = -\frac{1}{2}$, $\tau = e^{2\pi i/3}$, $\tau^{-1} = -\tau - 1$, $\pm \tau^* = (p + q\tau)/(1 + \tau) = q \pm \tau$, $q = 0$, $\tau^* = \tau$. This completes the proof of the sharpened condition.

10. The period parallelogram

Let ω_1, ω_2 be a basis of the period lattice with the imaginary part of the ratio $\tau = \omega_1/\omega_2$ positive. The periods ω_1, ω_2 determine in the complex w-plane a parallelogram $\mathfrak{P}$ whose points are $w = x\omega_1 + y\omega_2$, $0 \leq x < 1$, $0 \leq y < 1$; $\mathfrak{P}$ includes the sides issuing from 0 but not their end points ω_1 and ω_2. $\mathfrak{P}$ is called a *period parallelogram*. $\mathfrak{P}$ is not uniquely determined by the period lattice, since we can obtain other basic periods via arbitrary proper unimodular substitutions. To go around $\mathfrak{P}$ in the positive sense is to go around it so as to encounter the points 0, ω_1, $\omega_1 + \omega_2$, ω_2, in this order (Figure 23). All periods can be written in the form $\omega = m\omega_1 + n\omega_2$ with integer m and n, and different pairs m, n yield different periods. The translation defined by the vector ω carries $\mathfrak{P}$ into the parallelogram $\mathfrak{P}_{mn}$ with points $w = x\omega_1 + y\omega_2$, $m \leq x < m + 1$, $n \leq y < n + 1$. It is clear that every point of the w-plane belongs to precisely one such parallelogram. This decomposition of the plane into parallelograms generalizes the net of squares constructed in the preceding section.

We turn once more to the mapping defined by the meromorphic function $z = f(w)$. If C_1' is a curve in the w-plane joining 0 to w and C_1 is its image on $\mathfrak{R}$ joining $\mathfrak{z}_0$ to $\mathfrak{z}$, then, in view of Theorem 1 in Section 8, $\mathfrak{z}$ depends only on w and not on C_1'. If C_2 also joins $\mathfrak{z}_0$ to $\mathfrak{z}$, then $w(C_1 C_2^{-1}) = \omega$ is a period and $w = w(C_1) = w(C_2) + \omega$. On the other hand, given w, it is possible to find exactly one period ω such that the point $w - \omega$ lies in $\mathfrak{P}$. If we choose a closed curve A issuing from $\mathfrak{z}_0$ with $w(A) = \omega$ and put $C_2 = A^{-1}C_1$, then C_2 goes from $\mathfrak{z}_0$ to $\mathfrak{z}$, and $w(C_2)$ lies in $\mathfrak{P}$. This shows that the mapping $z = f(w)$ establishes a one-to-one correspondence between the parallelogram $\mathfrak{P}$ and the Riemann surface $\mathfrak{R}$. In particular, let C_1' and C_2' be the segments from 0 to ω_1 and ω_2, that is, the sides of $\mathfrak{P}$ issuing from 0. For their image curves C_1 and C_2 on $\mathfrak{R}$, we have $w(C_1) = \omega_1$, $w(C_2) = \omega_2$; that is, these curves share the double point $\mathfrak{z}_0$ but have no other double points or common points.

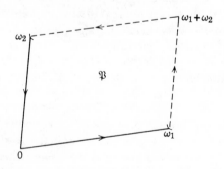

Figure 23

The periodicity of $f(w)$ implies that the segments from 0 to $-\omega_1$ and from 0 to $-\omega_2$ are mapped to C_1^{-1} and C_2^{-1}, and the conformal nature of the mapping implies that the curves C_2 and C_1 form at $_{3_0}$ the angle $\arg \tau$.

We now cut $\mathfrak{R}$ along the curves C_1 and C_2 and so produce two edges for each curve. Let $\mathfrak{R}_0$ denote the cut Riemann surface. Let $\mathfrak{P}$ now denote the closed period parallelogram (including all four sides). The function $z = f(w)$ maps $\mathfrak{P}$ conformally onto the surface $\mathfrak{R}_0$. In view of the periodicity of the mapping function, all other closed parallelograms $\mathfrak{P}_{mn}$ are also mapped onto $\mathfrak{R}_0$. We now take for each pair of indices m, n a copy $\mathfrak{R}_{mn}$ of $\mathfrak{R}_0$ and join the surfaces $\mathfrak{R}_{mn}$ along the edges of the cuts in just the way in which the $\mathfrak{P}_{mn}$ are joined in the network of parallelograms. The result is a certain Riemann surface. It is clear that the function $z = f(w)$ maps the w-plane conformally onto this Riemann surface, which is simply the covering surface $\mathfrak{S}$ introduced above.

It is now easy to determine how often the function $f(w)$ takes on a pre-assigned value z in the period parallelogram. The mapping $z = f(w)$ establishes a one-to-one correspondence $_3 \leftrightarrow w$ between $\mathfrak{R}$ and $\mathfrak{P}$. If z is not a branch point, then there lie over z exactly two distinct points of $\mathfrak{R}$ to which there correspond, in turn, two distinct points of $\mathfrak{P}$; here z may be the point at infinity. If z is one of the branch points a, b, c, d, then such a z leads to a single point $_3$ of $\mathfrak{R}$; but

$$\frac{df(w)}{dw} = \sqrt{P(z)}, \qquad \frac{d^2f(w)}{(dw)^2} = \frac{d\sqrt{P(z)}}{dz}\frac{dz}{dw} = \frac{1}{2}\frac{dP(z)}{dz},$$

$$P(z) = (z - a)(z - b)(z - c)(z - d),$$

implies that the function $f(w)$ takes on such a value z to the second order, and that precisely once in $\mathfrak{P}$. Incidentally, we see that if z is not a branch point then z is taken on twice to the first order. We see that, counting multiplicities, the function $f(w)$ takes on every preassigned value exactly twice in the period parallelogram.

We shall now introduce the concepts of a pair of crosscuts and of a canonical dissection which will be generalized in the sequel. Let C_1 and C_2 be two simple closed curves on $\mathfrak{R}$ issuing from the fixed point $_{3_0}$ which intersect at $_{3_0}$ and have no common points other than $_{3_0}$. Viewed as cuts on $\mathfrak{R}$, such curves are said to form a pair of *crosscuts* and to define a *canonical dissection* of the Riemann surface $\mathfrak{R}$. It is clear that in this definition the order of the curves C_1 and C_2 is irrelevant. So far we have encountered such canonical dissections twice: once in Section 5, when $\mathfrak{R}$ was mapped topologically onto a torus, which was subsequently dissected and transformed into a rectangle, and then, in the present section, when $\mathfrak{R}$ was mapped onto a period parallelogram. For our purposes, it is sufficient to assume that the curves C_1 and C_2 are piecewise smooth. We observe that the orientation of the sphere of complex numbers can be transferred to every Riemann surface, and this allows us to assume that the orientations of C_1 and C_2 are chosen so that C_1 crosses C_2 at $\mathfrak{z}_0$ from left to right in the manner of the real axis crossing the imaginary axis in the complex plane.

Let $\mathfrak{R}^*$ denote the result of canonical dissection of the Riemann surface along the crosscuts C_1 and C_2. We shall prove that the elliptic integral w maps $\mathfrak{R}^*$ conformally onto a parallelogram with, in general, curvilinear boundary. This was proved earlier in the special case when we started out with a parallelogram with rectilinear boundary, namely, the period parallelogram. If we start with an arbitrarily prescribed pair of crosscuts, then certain topological issues must be considered. We note, first of all, that $\mathfrak{R}^*$ remains connected. In fact, Figure 24 makes it clear that two opposite points on two edges resulting from dissection along C_2 can be connected in $\mathfrak{R}^*$. A similar statement holds with C_1 in place of C_2. Now, if we connect two points on the uncut surface $\mathfrak{R}$ by means of a curve, then we can dispense in the manner just indicated with all points of intersection with C_1 and C_2, and thus obtain a curve in $\mathfrak{R}^*$ joining our two points.

Let C_1' and C_2' again denote the images in the w-plane of C_1 and C_2. Both issue from $w = 0$ and have end points $w(C_1) = w_1$ and $w(C_2) = w_2$. Since C_1 and C_2 are simple closed curves, it follows from Theorem 1 in Section 8 that C_1' and C_2' have no double points and no common point different from the end points. We shall now show that the three numbers 0, w_1, w_2 are distinct. If all three numbers were equal, then C_1' and C_2' would be closed and, in view of the conformal nature of the mapping, would cross at 0. But this is impossible, for the simple closed curve C_1' separates the w-plane into two parts, whereas we could get from one side of C_1' to the other along C_2' without intersecting C_1' (Figure 25). This shows that w_1 and w_2 are not both zero. Suppose now that $w_1 = 0$ and $w_2 \neq 0$. Then at least C_1' would be a simple

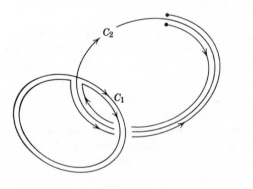

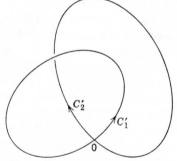

Figure 24   Figure 25

closed curve, and, reasoning as above, we could show that, apart from its origin 0, the curve C_2' would lie entirely in the interior or in the exterior of C_1'. The second of these possibilities can be reduced to the first by replacing C_1 and C_2 by C_1^{-1} and C_2^{-1}. But then, for a sufficiently large integer n, the point nw_2 would lie in the exterior of C_1' so that the curves C_1' and $(C_2')^n$ would have a first point of intersection in which $(C_2')^n$ intersects the curve C_1' from the inside to the outside (Figure 26). Since the curves C_2^n and C_1 on $\Re$ have in common only the point $\mathfrak{z}_0$, which is also the end point of the curves C_2^k ($k = 1, \ldots, n$), then, in view of Theorem 1, the above point of intersection of the curves C_1' and $(C_2^n)'$ would again be the point $w = 0$. This time we get a double contradiction, namely, the same contradiction as above and, in addition, for some natural number k, the false formula $kw_2 = 0$. A similar argument rules out the possibility $w_2 = 0$ and $w_1 \neq 0$. It follows that $w_1 \neq 0$ and $w_2 \neq 0$. Since nw_1 ($n = \pm 1, \pm 2, \ldots$) is the end point of the curve $(C_1^n)'$ we obtain for $n \to \pm \infty$ a simple curve C' which goes to infinity in both directions, is invariant under translation by w_1, and separates the w-plane into two parts. If $w_1 = w_2$, then C_1' and C_2' would bound a 2-gon, and would be sensed one way at 0 and the opposite way at w_1 (Figure 27). This would contradict the fact that C_1 and C_2 intersect in a certain definite sense at $\mathfrak{z}_0$. We conclude that $w_1 \neq w_2$. By applying a similar argument to C_1 and C_2^{-1}, we can show that $w_1 \neq -w_2$, that is, $w_1 + w_2 \neq 0$.

Now we consider the curve $C_1 C_2$ on $\Re$. This curve has exactly one multiple point on $\Re$, namely the point $\mathfrak{z}_0$, which is the beginning point of $C_1 C_2$, the end point of $C_1 C_2$, and the end point of the subarc C_1. To obtain the image $(C_1 C_2)'$ in the w-plane of the curve $C_1 C_2$, we adjoin to C_2' the curve obtained by subjecting the curve C_2' to the translation defined by the vector w_1. The multiple points of this image curve, if any, are among the images $0, w_1 + w_2, w_1$

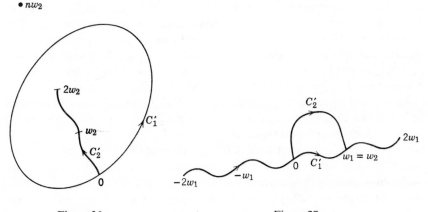

Figure 26   Figure 27

of $\mathfrak{z}_0$. But these three numbers are distinct, so that $(C_1C_2)'$ has no double points and the same is true for the image curve $(C_2C_1)'$ of C_2C_1. The image curves $(C_1C_2)'$ and $(C_2C_1)'$ have the common origin 0, and the common end point $w_1 + w_2$; at the same time the four relevant images 0, w_1, w_2, $w_1 + w_2$ w_1 of $\mathfrak{z}_0$ are distinct. If the two simple curves $(C_1C_2)'$ and $(C_2C_1)'$ had a common interior point, then this point would be the image of a point on C_1C_2 and C_2C_1 different from $\mathfrak{z}_0$ and would, therefore, belong to exactly one of the curves C_1 and C_2. In the first case, we would have $w_2 = 0$, and in the second case we would have $w_1 = 0$, and these possibilities are ruled out. We may therefore claim that the two simple curves $(C_1C_2)'$ and $(C_2C_1)'$ have in

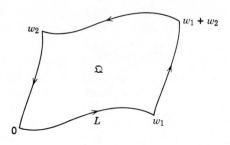

Figure 28

common only the beginning point and the end point. This proves that the image curve $L = (C_1C_2C_1^{-1}C_2^{-1})'$ of the boundary $C_1C_2C_1^{-1}C_2^{-1}$ of $\mathfrak{R}^*$ is a rectilinear or curvilinear parallelogram with vertices 0, w_1, $w_1 + w_2$, w_2, in that order (Figure 28).

We now show that the elliptic integral w maps $\mathfrak{R}^*$ onto the parallelogram-like region $\mathfrak{Q}$ with boundary L. We let w vary over the boundary L of $\mathfrak{Q}$ and consider the angle by which the direction of the tangent to L changes. The changes along each pair of opposite sides are equal and opposite. The mapping effected by $z = f(w)$ is conformal and orientation preserving, so that the changes in the direction of the tangent at the four vertices 0, w_1, $w_1 + w_2$, w_2 add up at $\mathfrak{z}_0$ to 2π. It follows that the angular increase involved in traversing L is 2π, and this indicates that $\mathfrak{Q}$ lies to the left of L. Now let C_0 be a closed curve on $\mathfrak{R}^*$ issuing from $\mathfrak{z}_0$ whose beginning and end tangents lie in the first quadrant at $\mathfrak{z}_0$ between C_1 and C_2. The image curve C_0' of C_0 in the w-plane enters the interior of $\mathfrak{Q}$ at the beginning point 0. If C_0' were not closed, that is, if $w(C_0) = \omega \neq 0$, then we could find a natural number n such that $n\omega$ is outside $\mathfrak{Q}$. But then, arguing as on a previous occasion, we could show that there is a point on the image curve $(C_0^n)'$ of C_0^n at which it crosses from the boundary of $\mathfrak{Q}$ into the exterior. Going back to C_0^n, we see that this curve and, therefore, also C_0, would have to cross the boundary of

$\mathfrak{R}^*$. This justifies the conclusion that $\omega = 0$. In other words, C_0' is closed and is entirely contained in $\mathfrak{Q}$ (Figure 29). Conversely, if C_0' is such a curve, then it is the image under the inverse mapping $z = f(w)$ of a curve which begins and ends in the first quadrant at $\mathfrak{z}_0$, and lies entirely in $\mathfrak{R}^*$. If C denotes a curve on $\mathfrak{R}^*$ with beginning point $\mathfrak{z}_0$ and end point $\mathfrak{z}$, then, as we have just proved, the value $w(C)$ of the elliptic integral depends on $\mathfrak{z}$ alone, and this implies that the analytic continuation of the function element $w(z)$ on $\mathfrak{R}^*$ is a single-valued function of $\mathfrak{z}$. This shows that the elliptic integral w effects a one-one conformal mapping of $\mathfrak{R}^*$ onto the region $\mathfrak{Q}$ as asserted. In view of the well-known fact that every plane region bounded by a simple

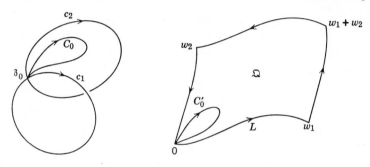

Figure 29

closed curve is simply connected, it follows that the dissected Riemann surface $\mathfrak{R}^*$ is simply connected.

Now we compute the area μ of $\mathfrak{Q}$. If we put $w = u + iv$, then

$$(1) \qquad \mu = \iint_{\mathfrak{Q}} du\ dv,$$

and by transforming (1) into a line integral, we have

$$\mu = \frac{1}{2}\int_L (u\ dv - v\ du),$$

where L is the positively oriented boundary of $\mathfrak{Q}$. Since

$$u\ du + v\ dv = d\left(\frac{u^2 + v^2}{2}\right)$$

is a total differential and

$$\bar{w}\ dw = (u - iv)(du + i\ dv) = (u\ du + v\ dv) + i(u\ dv - v\ du),$$

we have

$$2i\mu = \int_L \bar{w}\ dw.$$

Since the third side of $\mathfrak{Q}$ is the result of applying the translation w_2 to the first side, and the second side of $\mathfrak{Q}$ is the result of applying the translation w_1 to the fourth side, we conclude, bearing in mind orientations, that

$$2i\mu = \int_{C_1'} \{\bar{w} - (\bar{w} + \bar{w}_2)\}\, dw - \int_{C_2'} \{\bar{w} - (\bar{w} + \bar{w}_1)\}\, dw$$

$$= \bar{w}_1 \int_{C_2'} dw - \bar{w}_2 \int_{C_1'} dw,$$

so that

(2) $$\qquad\qquad\qquad 2i\mu = \bar{w}_1 w_2 - \bar{w}_2 w_1.$$

This result can be read off from Figure 30, which shows that the curvilinear parallelogram and the rectilinear parallelogram have the same areas. The advantage of the computational proof is that it applies to arbitrary rectifiable curves including curves whose complexity precludes visualization.

Let us now replace u and v in (1) by x and y, where $z = x + iy$ and $w = w(z)$. In view of the Cauchy-Riemann equations, we have

$$u_x = v_y, \qquad u_y = -v_x, \qquad \frac{dw(z)}{dz} = u_x + iv_x;$$

and this yields for the functional determinant $d(u, v)/d(x, y)$ the value

$$\frac{d(u, v)}{d(x, y)} = u_x v_y - v_x u_y = u_x^2 + v_x^2 = (u_x + iv_x)(u_x - iv_x)$$

$$= \left| \frac{dw}{dz} \right|^2 = |Q(z)|^2 = |P(z)|^{-1}.$$

Since the mapping of $\mathfrak{Q}$ onto $\mathfrak{R}^*$ is one-one, it follows that

(3) $$\qquad\qquad\qquad \mu = \iint_R \frac{dx\,dy}{|P(z)|},$$

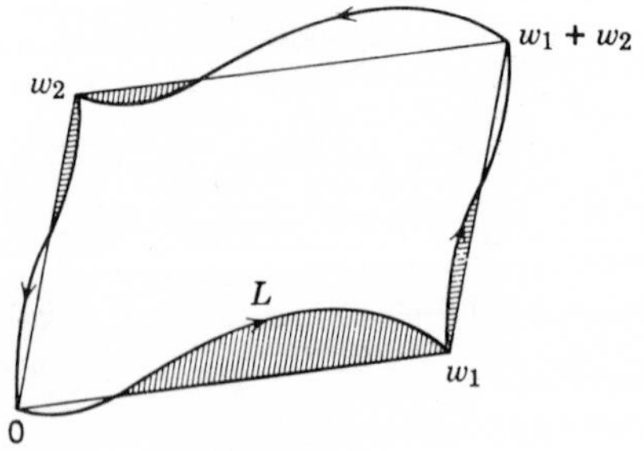

Figure 30

where it should be noted that integration over $\Re$ gives the same result as integration over $\Re^*$. This shows that the area μ of the curvilinear parallelogram Ω is independent of the canonical dissection. Since μ in (3) is positive, this argument provides an alternative proof of the assertion that the boundary of Ω, originally denoted by L, must be positively oriented. In view of (2), the imaginary part of the quotient w_2/w_1 is now seen to be positive, for it has the value $\mu|w_1|^{-2}$. Now consider the special canonical dissection defined by means of the rectilinear parallelogram $\mathfrak{P}$ with vertices 0, ω_1, $\omega_1 + \omega_2$, ω_2 introduced earlier. If we write the basic periods ω_1 and ω_2 in the form

$$w_1 = m\omega_1 + n\omega_2, \qquad w_2 = p\omega_1 + q\omega_2,$$

with integer m, n, p, q, then the invariance of the area μ implies the relation

$$mq - np = 1.$$

It follows that w_1 and w_2 are also a pair of basic periods, and so the pair of crosscuts yields a basis for the fundamental group Γ of $\Re$. Hence, if H is an arbitrary curve on $\Re$ and C is a fixed curve on $\Re$ both joining $\mathfrak{z}_0$ to a point $\mathfrak{z}$, then

$$H \sim C_1^k C_2^l C,$$

with integer exponents k and l uniquely determined by H. Two such curves H are homotopic if and only if the associated pair k, l is the same for both of them. For the elliptic integral we have

$$w(H) = w(C_1^k) + w(C_2^l) + w(C) = kw_1 + lw_2 + w(C).$$

Now let Ω_{kl} be the result of applying the translation (it is a period) $kw_1 + lw_2$ to Ω. Since the covering surface $\mathfrak{S}$ is mapped by the elliptic integral w in a one-to-one manner onto the full plane, it follows that, just as with the rectilinear parallelograms, the Ω_{kl} are linked in a gap-free manner and form a simple covering of the plane (Figure 31). It follows that $\mathfrak{S}$ can be put together by suitably linking infinitely many copies $\Re_{kl}^*$ ($k, l = 0, \pm 1, \ldots$) of $\Re^*$.

We note without proof that it is possible to deform the curvilinear parallelogram Ω continuously into a rectilinear parallelogram keeping its vertices fixed and admitting in the process of deformation only schlicht curvilinear parallelograms. Similarly, it is possible to deform the pairs of given crosscuts C_1, C_2 on $\Re$ continuously so that the images of the new crosscuts in the w-plane are segments, and the dissected Riemann surface is mapped onto a rectilinear period parallelogram. We do not, however, propose to go more deeply into this matter, for there are unsolved problems which appear when one attempts to generalize the relevant issues to Riemann surfaces of arbitrary algebraic functions.

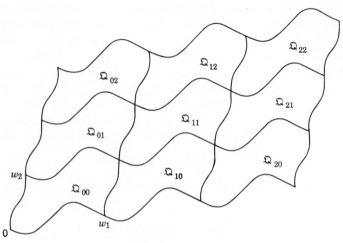

Figure 31

11. The $\wp$-function

Consider the elliptic integral

$$w = w(C) = \int_{z_0}^{z} Q(z)\, dz$$

over a path C on the Riemann surface $\mathfrak{R}$ from the point $\mathfrak{z}_0$ over z_0 to the point $\mathfrak{z}$ over z. The end points $\mathfrak{z}_0$ and $\mathfrak{z}$ of C determine w only up to an additive period. In addition, w depends on the four branch points a, b, c, d of the Riemann surface. The five parameters z_0, a, b, c, d can be reduced by means of a linear substitution in the manner described below.

The most general conformal mapping of the complex w-plane onto itself is defined by a transition from w to

$$(1) \qquad\qquad W = rw + t \qquad (r \neq 0)$$

with arbitrary constants r, t. The mapping properties established earlier for w and for the inverse function $z = f(w)$ carry over to the new function W. Translation by the vector t can be used to alter arbitrarily the beginning point $\mathfrak{z}_0$ of the path of integration C, and multiplication by r can be used to multiply the integrand $Q(z)$ by an arbitrary nonzero constant. Since

$$Q(z) = \frac{1}{\sqrt{P(z)}}, \qquad P(z) = (z-a)(z-b)(z-c)(z-d),$$

it follows that in this way we can obtain under the integral sign the reciprocal of the square root of an arbitrary polynomial of degree four with simple roots. Thus W becomes the general elliptic integral of the first kind as defined in Section 2. It is clear that the transition from w to $w + t$ has no effect on the period lattice; incidentally, the independence of the system of periods from the beginning point $\mathfrak{z}_0$ was established in Section 9. Multiplication by r has the effect of subjecting the period lattice to a dilation with coefficient $|r|$. Under this dilation the two basic periods ω_1 and ω_2 go over into the basic periods $r\omega_1$ and $r\omega_2$, and the ratio ω_2/ω_1 of the periods remains unchanged.

We now apply to the variable of integration z a fractional linear substitution

$$(2) \qquad z = \frac{\alpha s + \beta}{\gamma s + \delta} \qquad (\alpha\delta - \beta\gamma \neq 0),$$

with constant α, β, γ, δ, and show as in Section 2 that under such a substitution an elliptic integral of the first kind goes over into an elliptic integral of the first kind. In fact,

$$\frac{dz}{ds} = \frac{\alpha\delta - \beta\gamma}{(\gamma s + \delta)^2}, \qquad P(z) = \left(\frac{\alpha s + \beta}{\gamma s + \delta} - a\right) \cdots \left(\frac{\alpha s + \beta}{\gamma s + \delta} - d\right) = \frac{P_1(s)}{(\gamma s + \delta)^4}$$

$$\frac{dz}{\sqrt{P(z)}} = (\alpha\delta - \beta\gamma)\frac{ds}{\sqrt{P_1(s)}},$$

where $P_1(s)$ is a polynomial of degree four or three in s. The zeros of $P_1(s)$ are obtained from the numbers a, b, c, d through the inverse substitution

$$s = \frac{\delta z - \beta}{\alpha - \gamma z},$$

provided we agree not to count the possible value ∞ as a zero. Since the linear substitution (2) does not change when we multiply α, β, γ, δ by the same nonzero factor, the two substitutions (1) and (2) give us five parameters, namely, r, t and the three ratios $\alpha/\beta/\gamma/\delta$. This enables us to adjust suitably the five parameters, namely, the lower limit z_0 and four of the five coefficients of the polynomial of degree four which appear in the general elliptic integral of the first kind. In this way it is possible to obtain the *Legendre normal form*

$$w = \int_0^s \frac{d\sigma}{\sqrt{(1 - \sigma^2)(1 - k^2\sigma^2)}},$$

where k is the only free parameter of the integral. Since the zeros of the polynomial are ± 1, $\pm k^{-1}$, it follows that $k \neq 0$, ± 1.

Instead of the Legendre normal form we shall discuss in some detail the Weierstrass normal form. To simplify computations we write (2) as a product of two substitutions. The first substitution is

$$z = a + \frac{1}{s_1}, \qquad dz = -\frac{ds_1}{s_1^2}.$$

This substitution shifts the branch point a to ∞; we are assuming $z_0 = a$. Hence

$$w = \int_\infty^{s_1} \frac{d\sigma}{\sqrt{P_1(\sigma)}},$$

where

$$P_1(\sigma) = a_0\sigma^3 + a_1\sigma^2 + \cdots, \qquad a_0 = (a-b)(a-c)(a-d) \neq 0.$$

The second substitution is

$$s_1 = \gamma s + \delta \qquad (\gamma \neq 0)$$

with γ, δ constant. As a result of this substitution we have

$$w = \int_\infty^{s} \frac{d\sigma}{\sqrt{P_2(\sigma)}}, \qquad P_2(\sigma) = q_0\sigma^3 + q_1\sigma^2 + \cdots = \gamma^{-2}P_1(\gamma\sigma + \delta),$$

where

$$q_0 = a_0\gamma, \qquad q_1 = 3a_0\delta + a_1.$$

If we choose

$$\gamma = \frac{4}{a_0}, \qquad \delta = -\frac{a_1}{3a_0},$$

then $q_0 = 4$, $q_1 = 0$ and

$$P_2(\sigma) = 4\sigma^3 - g_2\sigma - g_3,$$

with certain constants g_2, g_3. In this way we arrive at the *Weierstrass normal form*

$$(3) \qquad\qquad w = \int_\infty^{s} \frac{d\sigma}{\sqrt{4\sigma^3 - g_2\sigma - g_3}}.$$

Let e_1, e_2, e_3 denote the three zeros of the cubic polynomial $P_2(\sigma)$. Since the quadratic term is missing we have

$$e_1 + e_2 + e_3 = 0.$$

The fact that the three roots are distinct is expressed most simply by forming the discriminant

$$g_2^3 - 27g_3^2 = 16(e_1 - e_2)^2(e_2 - e_3)^2(e_3 - e_1)^2 \neq 0.$$

The two substitutions used to derive the Weierstrass normal form yield the following single fractional linear substitution,

$$(4) \qquad z = a + \frac{1}{\gamma s + \delta} = \frac{(a\gamma)s + (a\delta + 1)}{\gamma s + \delta} = \frac{\alpha s + \beta}{\gamma s + \delta}$$

with determinant

$$\alpha\delta - \beta\gamma = a\gamma\delta - (a\delta + 1)\gamma = -\gamma = \frac{4}{(b-a)(c-a)(d-a)} \neq 0.$$

If we solve (4) for s and note that $z = f(w)$, then we see that s is also a function of w. We call this function the $\wp$-*function*; in other words, the $\wp$-function is the inverse function of the elliptic integral of the first kind in the Weierstrass normal form. We have

$$s = \wp(w) = \frac{\delta f(w) - \beta}{\alpha - \gamma f(w)}.$$

Since $f(w)$ is a doubly periodic meromorphic function, the same is true of $\wp(w)$. Again, since in the period parallelogram the function $f(w)$ takes on every value (counting multiplicities) exactly twice, the same is true of the $\wp$-function.

Theorem 1: The $\wp$-function has in the period parallelogram exactly one pole, namely the double pole at $w = 0$.

Proof: To $s = \infty$ there corresponds via the linear mapping (4) the branch point $z = a$. We saw in the previous section that in the period parallelogram $w = \xi\omega_1 + \eta\omega_2$ ($0 \leq \xi < 1, 0 \leq \eta < 1$), the function $f(w)$ takes on the branch point $z = a$ exactly once to the second order. On the other hand, (3) shows that $w = 0$ at $s = \infty$ when we shrink the path of integration to the point $s = \infty$. This means that $w = 0$ is the required pole of the $\wp$-function.

That $w = 0$ is a double pole of the $\wp$-function can be deduced from the differential equation implied by (3). The differential equation in question is

$$\left(\frac{dw}{ds}\right)^2 = (4s^3 - g_2 s - g_3)^{-1}.$$

It follows that the inverse function $\wp(w)$ satisfies the differential equation

(5)
$$\left(\frac{d\wp(w)}{dw}\right)^2 = 4(\wp(w))^3 - g_2\wp(w) - g_3.$$

Let

$$\wp(w) = \frac{c_{-n}}{w^n} + \cdots + \frac{c_{-1}}{w} + c_0 + c_1 w + \cdots (c_{-n} \neq 0)$$

be the Laurent expansion of the $\wp$-function at the pole $w = 0$. We know the $n \geq 1$. This implies

$$(\wp(w))^3 = \left(\frac{c_{-n}}{w^n}\right)^3 + \cdots, \qquad \left(\frac{d\wp(w)}{dw}\right)^2 = \left(\frac{-nc_{-n}}{w^{n+1}}\right)^2 + \cdots.$$

Equating coefficients, we find that
$$3n = 2n + 2, \qquad (-nc_{-n})^2 = 4(c_{-n})^3,$$

that is, $n = 2$ and $c_{-2} = 1$. Before determining the other coefficients of the Laurent expansion we prove the following simple property of the $\wp$-function.

Theorem 2: The function $\wp(w)$ is even.

Proof: The lower limit ∞ of the elliptic integral (3) is a branch point of the Riemann surface $\Re$ of the integrand. Since the two sheets of $\Re$ are joined at a branch point, we see that to every curve C on $\Re$ issuing from the branch point in question there corresponds a curve C^* on $\Re$ such that C and C^* have the same projection on the s-sphere. It follows that the values of the integrand at two points of $\Re$ which lie one above the other are equal and opposite. In this way to every projected point s there corresponds in addition to the value w the value $-w$ which is an admissible value of the integral obtained by replacing C with C^*. This means that for the inverse function we have

$$\wp(w) = \wp(-w),$$

which proves our assertion. A different proof goes back to Section 7 in which we expanded the inverse function $z = f(w)$ in a power series in the neighborhood of the image w_1 of a branch point z_1. This power series contains only even powers of $w - w_1$. In particular, taking $z_1 = a$ and $w_1 = 0$ we see that $z = f(w)$ is an even function; but then $s = \wp(w)$ is also an even function. True, at first our conclusion is valid only locally in the neighborhood of $w = 0$, but then it can be justified in general using analytic continuation.

In view of Theorem 2 the Laurent series for $\wp(w)$ takes the simpler form

$$\wp(w) = w^{-2} + b_0 + b_1 w^2 + b_2 w^4 + \cdots.$$

Therefore,

$$(\wp(w))^3 = w^{-6}(1 + b_0 w^2 + b_1 w^4 + b_2 w^6 + \cdots)^3,$$

$$\left(\frac{d\wp(w)}{dw}\right)^2 = w^{-6}(-2 + 2b_1 w^4 + 4b_2 w^6 + \cdots)^2.$$

Equating the coefficients of w^{-4} in (5), we get $12b_0 = 0$; that is, $b_0 = 0$. Equating the coefficients of w^{-2} and the constant terms, we get

$$-8b_1 = 12b_1 - g_2, \qquad -16b_2 = 12b_2 - g_3;$$

that is,

(6) $$b_1 = \frac{g_2}{20}, \qquad b_2 = \frac{g_3}{28}.$$

For greater ease in obtaining the coefficients $b_3, b_4, \ldots,$ we do not apply the technique of equating coefficients to (5). Instead, we differentiate our differential equation. The result is

$$2\wp'\wp'' = 12\wp^2\wp' - g_2\wp'.$$

Since $\wp'(w)$ is not identically zero, we see that

$$\wp''(w) = 6(\wp(w))^2 - \frac{g_2}{2}.$$

In view of the form of our Laurent series, this equation leads to the identity

$$6w^{-4} + \sum_{n=1}^{\infty} 2n(2n-1)b_n w^{2n-2} = 6w^{-4}\left(1 + \sum_{n=1}^{\infty} b_n w^{2n+2}\right)^2 - \frac{g_2}{2}.$$

Equating the coefficients of w^{2n-2} for $n > 2$ on both sides, we obtain

$$2n(2n-1)b_n = 6(2b_n + b_1 b_{n-2} + b_2 b_{n-3} + \cdots + b_{n-2}b_1).$$

Since

$$2n(2n-1) - 12 = 2(2n+3)(n-2) \neq 0,$$

we obtain a recursive formula for the coefficients b_n with $n > 2$.

Theorem 3:   The Laurent expansion of the $\wp$-function at 0 has the form

$$\wp(w) = w^{-2} + \sum_{n=1}^{\infty} b_n w^{2n}$$

with

$$b_1 = \frac{g_2}{20}, \qquad b_2 = \frac{g_3}{28}, \qquad b_n = \frac{3}{(2n+3)(n-2)} \sum_{k=1}^{n-2} b_k b_{n-k-1}$$

$$(n = 3, 4, \ldots).$$

The recursion formula shows that all the coefficients b_n ($n = 1, 2, \ldots$) are polynomials in g_2, g_3 with positive rational coefficients, for example;

$$b_3 = \frac{b_1^2}{3} = \frac{g_2^2}{1200}.$$

The Laurent series converges exactly in the largest circle about the pole $w = 0$, which contains no other singularities of the function in its interior. Since the $\wp$-function has in the period parallelogram only the pole $w = 0$, it follows from the periodicity that the poles of the $\wp$-function are precisely the lattice points of the period lattice. Hence the boundary of the circle of convergence of the Laurent series passes through the lattice point $\omega_1 \neq 0$ which is closest to zero, and the radius of convergence is $|\omega_1|$.

12. Partial fractions expansion of the $\wp$-functions

Let ω_1 and ω_2 be two distinct nonzero numbers whose ratio ω_2/ω_1 is not real. The numbers ω_1 and ω_2 may be the two basic periods considered above, but this assumption is not immediately necessary. We form the numbers

$$\omega = m\omega_1 + n\omega_2 \qquad (m, n = 0, \pm 1, \ldots),$$

where, as indicated, m and n are arbitrary integers, and then we form the infinite series

$$S = \sum_\omega |w - \omega|^{-\rho} \qquad (\rho > 2);$$

here w is a complex variable and the real number $\rho > 2$. For purposes of summation we suppose that the countably many numbers ω have been arranged in some definite order. If w is equal to an ω, then the corresponding term of the series S is omitted.

Theorem 1: For every fixed $\rho > 2$ the series S converges uniformly in every bounded region of the w-plane.

Proof: It suffices to prove uniform convergence in the disk $|w| < R$ for an arbitrary $R > 1$. Let

$$R_k = 2^k R \qquad (k = 0, 1, \ldots),$$

and let

$$\omega = m\omega_1 + n\omega_2 = (m + n\tau)\omega_1$$

be a lattice point lying in the disk $|w| < R_k$. Then

$$|m + n\tau| = |m + n\bar{\tau}| < R_k|\omega_1|^{-1},$$

so that

$$n|\tau - \bar{\tau}| = |(m + n\tau) - (m + n\bar{\tau})| < 2R_k|\omega_1|^{-1}.$$

It follows that the two ratios m/R_k and n/R_k lie between bounds which are independent of k, and so the number of the ω in every disk of radius R_k is less than $c\,R_k^2$, where c is a positive number independent of k. If we put

$$S_k = \sum_{R_{k-1} \leq |\omega| < R_k} |w - \omega|^{-\rho} \qquad (k = 2, 3, \ldots),$$

then, clearly, it suffices to prove the uniform convergence of the series

$$S^* = S_2 + S_3 + \cdots$$

with respect to w in the disk $|w| < R$. For $|\omega| \geq R_{k-1}$ we have

$$|w - \omega| \geq |\omega| - |w| > R_{k-1} - R \geq R_{k-2} \qquad (k = 2, 3, \ldots),$$

so that

$$S_k < cR_k^2 R_{k-2}^{-\rho} = cR^{2-\rho}2^{2\rho+k(2-\rho)}.$$

We see that S^* is majorized by a geometric progression free of w with quotient $2^{2-\rho} < 1$, and this proves our assertion.

A similar argument shows that the series S diverges for $\rho = 2$ and all the more so for $\rho < 2$.

Theorem 2: For the derivative of the $\wp$-function we have the partial fraction decomposition

$$\wp'(w) = -2 \sum_{\omega} (w - \omega)^{-3},$$

where we sum over all the periods ω of the $\wp$-function taken in arbitrary order.

Proof: The function $-\frac{1}{2}\wp'(w)$ is meromorphic and its poles are precisely the periods ω. To determine the principal parts at these poles we make use of the Laurent expansion obtained in the previous section,

$$\wp(w) = w^{-2} + b_1 w^2 + \cdots,$$

from which it follows that

$$-\tfrac{1}{2}\wp'(w) = w^{-3} - b_1 w + \cdots.$$

We see that the function $-\frac{1}{2}\wp'(w)$ has at 0 the principal part w^{-3}. This fact and the periodicity imply that the corresponding principal part at a pole ω is $(w - \omega)^{-3}$. In view of Theorem 1, the series

$$g(w) = \sum_{\omega} (w - \omega)^{-3}$$

made up of these principal parts converges uniformly in every bounded portion of the w-plane provided that we omit terms with $w = \omega$. In view of the Mittag Leffler theorem the function

$$g(w) + \tfrac{1}{2}\wp'(w) = h(w)$$

is entire. Observe that if ω_0 is any period of the $\wp$ function, and if ω ranges over all the periods then so does, in a different order, $\omega - \omega_0$. But the uniform convergence of the series $g(w)$ allows us to ignore the order of the terms and so assert that

$$g(w + \omega_0) = g(w).$$

We see that $g(w)$ has the same periods ω as $\wp(w)$ and $\wp'(w)$ and the same is therefore true of the entire function $h(w)$. Since $h(w)$ is bounded in the period parallelogram and periodic, it is bounded in the whole w plane and, therefore, in view of Liouville's theorem, constant. As ω ranges over all the periods so does $-\omega$. It follows that

$$g(-w) = \sum_{\omega} (-w + \omega)^{-3} = -g(w),$$

that is, $g(w)$ is an odd function. The same is true, by Theorem 2 in the previous section, of the function $\wp'(w)$. But then $h(w)$ is also odd and, therefore, identically equal to zero. This proves our theorem.

The fact proved in Theorem 1 allows us to dispense with convergence-producing summands in the partial fraction decomposition of $\wp'(w)$.

Theorem 3: The $\wp$-function has the partial fraction decomposition

$$\wp(w) = w^{-2} + \sum_{\omega \neq 0} \{(w - \omega)^{-2} - \omega^{-2}\},$$

where the sum is taken over all nonzero periods ω in any order whatsoever.

Proof: In view of Theorem 2 we have

$$(1) \qquad\qquad \wp'(w) + 2w^{-3} = -2 \sum_{\omega \neq 0} (w - \omega)^{-3}.$$

We integrate this relation along a path joining 0 to an arbitrary point w. Should this path run through lattice points ω other than 0, then the finite number of terms on the right side corresponding to those lattice points are shifted to the left side, where they vanish as a result of cancellation against the appropriate principal parts of the function $\wp'(w)$. The uniform convergence of the series which follows from Theorem 1 permits termwise integration and we have

$$[\wp(w) - w^{-2}]_0^w = \sum_{\omega \neq 0} [(w - \omega)^{-2}]_0^w.$$

In view of Theorem 3 in the previous section, the function $\wp(w) - w^{-2}$ vanishes at $w = 0$. We therefore obtain the formula

$$\wp(w) - w^{-2} = \sum_{\omega \neq 0} \{(w - \omega)^{-2} - \omega^{-2}\}.$$

This proves our theorem.

In the partial fraction decomposition there appear the convergence-producing summands $-\omega^{-2}$. That these summands are indispensable is obvious, for without them the series would no longer be absolutely convergent. There is clearly a certain analogy between the partial fraction decomposition of the $\wp$-function and the partial fraction decomposition of the cotangent function,

$$\operatorname{ctg} w = w^{-1} + \sum_{n \neq 0} \{(w - n\pi)^{-1} + (n\pi)^{-1}\}.$$

Here $n\pi$, $n = \pm 1, \pm 2, \ldots$, ranges over all the nonzero periods of the simply periodic function $\operatorname{ctg} w$.

As another application of Theorem 2, we shall express the coefficients b_n ($n = 1, 2, \ldots$) in the Laurent series for $\wp(w)$, obtained in the previous section, in terms of the periods ω. Differentiation of that Laurent series yields

$$\wp'(w) + 2w^{-3} = \sum_{n=1}^{\infty} 2nb_n w^{2n-1},$$

and for this function we have the partial fraction decomposition (1). In view of the uniform convergence of the series of partial fractions, the Weierstrass convergence theorem is applicable, and we can obtain the derivatives of the function through termwise differentiation. Putting $w = 0$ in the derivative of order $2n - 1$, we obtain the relation

$$(2n)! \, b_n = (2n + 1)! \sum_{\omega \neq 0} \omega^{-2n-2} \qquad (n = 1, 2, \ldots),$$

so that

(2) $$b_{n-1} = (2n - 1)\sigma_n \qquad (n = 2, 3, \ldots),$$

where, for $n = 2, 3, \ldots$,

$$\sigma_n = \sum_{\omega \neq 0} \omega^{-2n}$$

denotes the sum of the reciprocals of the $2n$-th powers of the nonzero period of the $\wp$-function. According to Theorem 3 of the previous section,

$$b_1 = \frac{g_2}{20}, \qquad b_2 = \frac{g_3}{28}, \qquad (2n + 3)(n - 2)b_n = 3 \sum_{k=1}^{n-2} b_k b_{n-k-1}$$

$$(n = 3, 4, \ldots).$$

In view of (2),

(3) $$\sum_{\omega \neq 0} \omega^{-4} = \sigma_2 = \frac{b_1}{3} = \frac{g_2}{60},$$

(4) $$\sum_{\omega \neq 0} \omega^{-6} = \sigma_3 = \frac{b_2}{5} = \frac{g_3}{140}$$

and, in general,

$$\sum_{\omega \neq 0} \omega^{-2n} = \sigma_n = \frac{b_{n-1}}{2n - 1} = P_n(\sigma_2, \sigma_3) \qquad (n = 2, 3, \ldots),$$

where $P_n(\sigma_2, \sigma_3)$ is a polynomial in σ_2, σ_3 with positive rational coefficients. This relation between σ_n, σ_2, σ_3 has, so far, never been proved without the use of function-theoretical tools. If we introduce two basic periods ω_1 and ω_2, then the formulas (3) and (4) can be given the form

$$g_2 = 60 \sum{}'(m\omega_1 + n\omega_2)^{-4}, \qquad g_3 = 140 \sum{}'(m\omega_1 + n\omega_2)^{-6},$$

where the summations extend over all pairs of integers m, n other than 0, 0. This shows, in particular, that the quantities g_2 and g_3 are uniquely determined by a pair of basic periods ω_1 and ω_2.

13. The inversion problem

So far we have defined the function $\wp(w)$ as the inverse of the elliptic integral of the first kind in the Weierstrass normal form

$$(1) \qquad w = \int_\infty^z \frac{d\zeta}{\sqrt{4\zeta^3 - g_2\zeta - g_3}},$$

with g_2 and g_3 subject to the single restriction $g_2^3 - 27g_3^2 \neq 0$. We showed that the $\wp$-function is a doubly periodic meromorphic function; specifically, we obtained two basic periods ω_1 and ω_2 as values of the integral corresponding to two curves C_1 and C_2 of a canonical dissection of the Riemann surface of $\sqrt{4z^3 - g_2z - g_3}$. The ratio ω_2/ω_1 is not real, and every period is uniquely represented in the form $\omega = m\,\omega_1 + n\,\omega_2$ with m and n arbitrary integers.

The inversion problem is concerned with the question of whether we can prescribe the numbers ω_1 and ω_2 arbitrarily except for the restriction that the ratio ω_2/ω_1 is not real. This means that we must find two numbers g_2 and g_3 such that $g_2^3 - 27g_3^2 \neq 0$ and such that the prescribed numbers ω_1 and ω_2 are basic periods of the elliptic integral (1) formed with these g_2 and g_3. The remark at the end of the previous section settles the uniqueness, if not the existence, of the solution g_2, g_3 of the inversion problem. The following two theorems provide a complete solution of the inversion problem.

Theorem 1: Let ω_1 and ω_2 be given numbers whose ratio is not real. If we put

$$p(w) = w^{-2} + \sum_{\omega \neq 0} \{(w - \omega)^{-2} - \omega^{-2}\},$$

where the summation extends over all $\omega = m\,\omega_1 + n\,\omega_2 \neq 0$ with integer m and n, then $p(w)$ is a doubly periodic meromorphic function which has precisely the ω as its nonzero periods.

Theorem 2: Further, if we put

$$(2) \qquad g_2 = 60 \sum_{\omega \neq 0} \omega^{-4}, \qquad g_3 = 140 \sum_{\omega \neq 0} \omega^{-6},$$

then $g_2^3 - 27g_3^2 \neq 0$, and the inverse function $\wp(w)$ of the elliptic integral (1) formed with these g_2, g_3 coincides with $p(w)$.

Proof of Theorem 1: The Weierstrass convergence theorem and Theorem 1 of the preceding section imply that the series

$$q(w) = -2 \sum_{\omega} (w - \omega)^{-3}$$

defines a meromorphic function. The poles of this function are precisely the points $w = \omega = m\,\omega_1 + n\,\omega_2$. It follows that ω are the only possible periods of this function. Since the ω are indeed periods of this function, they are all the periods of this function. Integrating over a path which avoids the periods, we get

$$(3) \qquad \int_0^w [q(w) + 2w^{-3}]\,dw = \sum_{\omega \neq 0} [(w - \omega)^{-2} - \omega^{-2}]$$

$$= p(w) - w^{-2},$$

so that

$$q(w) = \frac{dp(w)}{dw}.$$

As ω ranges over the periods so does $-\omega$. It follows that $p(w) = p(-w)$ is an even meromorphic function whose poles are precisely the points $w = \omega$ including 0.

It remains to show that the periods of $p(w)$ are precisely the numbers ω. The relation

$$\frac{dp(w)}{dw} = q(w) = q(w + \omega) = \frac{dp(w + \omega)}{dw}.$$

implies that

$$(4) \qquad p(w + \omega) = p(w) + \gamma,$$

where γ is independent of w but possibly dependent on ω. To prove that γ vanishes for all $\omega = m\,\omega_1 + n\,\omega_2$, it suffices to investigate the two cases $\omega = \omega_1$ and $\omega = \omega_2$. For $\omega = \omega_1$ or ω_2 the value $\omega/2$ is not a pole of the function $p(w)$. Putting $w = -\omega/2$ in (4), and bearing in mind the fact the $p(w)$ is even, we obtain $\gamma = 0$. It follows that all the ω are periods of $p(w)$. Since the origin is a pole, every period yields a pole. On the other hand, there are no poles other than the ω. This proves Theorem 1.

Proof of Theorem 2: Next we show that the function $p(w)$ satisfies a differential equation of the form

$$(5) \qquad \left(\frac{dp}{dw}\right)^2 = 4p^3 - g_2 p - g_3,$$

with suitably chosen coefficients g_2 and g_3. The relation (3) shows that the function $p(w) - w^{-2}$ is regular at $w = 0$ and has there the value 0. Since it is also even, its power series expansion about the origin has the form

$$(6) \qquad p(w) = w^{-2} + \sum_{n=1}^{\infty} c_n w^{2n},$$

and its circle of convergence passes through a period $\omega \neq 0$ closest to the origin. Differentiation yields

$$\frac{dp(w)}{dw} = -2w^{-3} + 2\sum_{n=1}^{\infty} nc_n w^{2n-1}.$$

Now we form the series

$$\left(\frac{dp(w)}{dw}\right)^2 = 4w^{-6} - 8c_1 w^{-2} - 16c_2 + \cdots,$$

$$4[p(w)]^3 = 4w^{-6} + 12c_1 w^{-2} + 12c_2 + \cdots,$$

$$g_2 p(w) + g_3 = g_2 w^{-2} + g_3 + \cdots,$$

where the positive powers of w have been left out. If we put

$$(7) \qquad g_2 = 20c_1, \qquad g_3 = 28c_2,$$

then the function

$$h(w) = \left[\frac{dp(w)}{dw}\right]^2 - 4[p(w)]^3 + g_2 p(w) + g_3$$

is regular at $w = 0$ and vanishes there. Like $p(w)$ the function $h(w)$ has all the ω as periods and is, therefore, regular at $w = \omega$. On the other hand, $h(w)$ is a meromorphic function which cannot have poles other than $p(w)$. Since the poles of $p(w)$ are precisely the ω, we conclude that $h(w)$ is an entire function. $h(w)$ is bounded in the period parallelogram spanned by ω_1 and ω_2 and, therefore, because of its periodicity, in the whole w-plane. But then in view of Liouville's theorem, $h(w)$ is constant; in fact, $h(w)$ is identically zero since this is its value at the origin. This proves that $p = p(w)$ satisfies the differential equation (5) with g_2 and g_3 defined by (6) and (7). The equations (2) for g_2 and g_3 follow from the considerations at the end of previous section.

We now show that $g_2^3 - 27g_3^2 \neq 0$. If e_1, e_2, e_3, are the zeros of the cubic polynomial

$$4p^3 - g_2 p - g_3 = 4(p - e_1)(p - e_2)(p - e_3),$$

then we must show that these three numbers are distinct. To this end we make use of a result which we shall establish in the next section independently of the present considerations. Specifically, we shall prove in the next section that for every complex number c, the equation $p(w) = c$ has a solution w.

Let m be the multiplicity of the zero e_1, and $w = w_1$ a solution of the equation $p(w) = e_1$ of multiplicity n. Then the function $[dp(w)/dw]^2$ vanishes at w_1 to order $2n - 2$; on the other hand, the cubic polynomial contains the factor $(p - e_1)^m$, which for $p = p(w)$ as a function of w vanishes at w_1 to order mn. It now follows from the differential equation (5) that

$$2n - 2 = mn, \qquad (2 - m)n = 2, \qquad m = 1, \qquad n = 2.$$

We conclude that the zeros e_1, e_2, e_3 are simple, and so we can form the elliptic integral (1) whose inverse function $z = \wp(w)$ is the Weierstrass $\wp$-function which also satisfies the differential equation (5).

It remains to show that $p(w)$ and $\wp(w)$ are the same function. For this it suffices to show that the two Laurent series

$$p(w) = w^{-2} + \sum_{n=1}^{\infty} c_n w^{2n}, \qquad \wp(w) = w^{-2} + \sum_{n=1}^{\infty} b_n w^{2n}$$

are identical. Observe that both functions $p(w)$ and $\wp(w)$ satisfy the same differential equation (5), are even, and have a pole at $w = 0$. As shown at the end of Section 11 this information determines a unique Laurent expansion about $w = 0$. Hence $p(w) = \wp(w)$ as asserted.

The uniqueness theorem for the solutions of the differential equation (5) of which we have just made use can be employed to obtain a proof of the inequality $g_2^3 - 27g_3^2 \neq 0$ which is independent of the next section. In fact, suppose we had $e_1 = e_2$. Then $e_1 + e_2 + e_3 = 0$ implies $e_3 = -2e_1$. If $e_1 = 0$, then $g_2 = g_3 = 0$, and the corresponding differential equation

$$\left(\frac{dq}{dw}\right)^2 = 4q^3$$

is satisfied by the function

$$q = q(w) = w^{-2}.$$

If $e_1 \neq 0$, we define

$$a = \sqrt{-3e_1}, \qquad q(w) = e_1 + a^2[\sin (aw)]^{-2}.$$

The corresponding differential equation for $q(w)$ is

$$\left(\frac{dq}{dw}\right)^2 = 4(q - e_1)^2(q + 2e_1),$$

with

$$g_2 = 12e_1^2 = \tfrac{4}{3}a^4, \qquad g_3 = -8e_1^3 = \tfrac{8}{27}a^6.$$

In both cases $q(w)$ is an even meromorphic function with pole at $w = 0$, but not a doubly periodic function; and yet, in view of the above uniqueness theorem we must have $q(w) = p(w)$.

14. The field of elliptic functions

Let ω_1 and ω_2 again be two numbers whose ratio is not real. By possibly reversing the order of ω_1 and ω_2, we can assume that the imaginary part of the ratio ω_2/ω_1 is positive. The points $\omega = m\omega_1 + n\omega_2$ with arbitrary integer m, n form in the complex plane a lattice which, in turn, determines the basis ω_1, ω_2 up to a proper unimodular substitution. In the sequel we shall assume the choice of a definite basis. The complex variable w will from now on be denoted by z.

We shall call a function $E(z)$ *elliptic* if it is meromorphic and satisfies the two relations

$$E(z + \omega_1) = E(z), \qquad E(z + \omega_2) = E(z)$$

for all z. It is clear that all the ω are periods of $E(z)$, and we admit the possibility that $E(z)$ has additional periods. Every constant is, trivially, an elliptic function. If $E_1(z)$ and $E_2(z)$ are two elliptic functions, then their sum $E_1(z) + E_2(z)$, their difference $E_1(z) - E_2(z)$, their product $E_1(z)E_2(z)$ and, if $E_2(z)$ is not identically zero, their quotient $E_1(z)/E_2(z)$ are all meromorphic functions with the periods ω. Hence the elliptic functions associated with the prescribed basis ω_1, ω_2 form a field. This field contains the complex numbers, but it also contains nonconstant functions, such as the function

$$\wp(z) = z^{-2} + \sum_{\omega \neq 0} [(z - \omega)^{-2} - \omega^{-2}],$$

whose periods are precisely the ω. The derivative

$$\frac{d\wp(z)}{dz} = \wp'(z) = -2 \sum_{\omega} (z - \omega)^{-3}$$

of $\wp(z)$ is also an elliptic function. More generally, every quotient

$$E(z) = \frac{F(\wp, \wp')}{G(\wp, \wp')}, \qquad [\wp = \wp(z), \wp' = \wp'(z)],$$

where F and G are polynomials in $\wp, \wp'$ and $G(\wp, \wp')$ does not vanish identically in z, is an elliptic function. We shall show in the sequel that every elliptic function is representable as the quotient of two polynomials in $\wp, \wp'$ with constant coefficients.

In view of their periodicity, it suffices to investigate the elliptic functions in the period parallelogram $\mathfrak{P}$ defined by

$$z = \xi\omega_1 + \eta\omega_2 \qquad (0 \leq \xi < 1, 0 \leq \eta < 1).$$

We shall denote the positively oriented boundary of $\mathfrak{P}$ by L, the side from 0 to ω_1 by C_1, and the subsequent sides by C_2, C_3, C_4. We note that C_2 and C_3

do not belong to $\mathfrak{P}$, and that of the end points of C_1 and C_4 only the end point 0 belongs to $\mathfrak{P}$.

We recall that the residue of a meromorphic function at a pole z_1 is the coefficient of $(z - z_1)^{-1}$ of its Laurent expansion about z_1. The sum of the residues of an elliptic function is defined as the sum of the residues at its poles in $\mathfrak{P}$.

Theorem 1: The sum of the residues of an elliptic function is 0.

Proof: If an elliptic function $E(z)$ has no pole on the boundary L of $\mathfrak{P}$, then, by the residue theorem the sum of its residues is

$$r = \frac{1}{2\pi i} \int_L E(z)\, dz.$$

In view of the periodicity of $E(z)$, we have

$$\int_{C_1} E(z)\, dz = -\int_{C_3} E(z)\, dz \quad \text{and} \quad \int_{C_2} E(z)\, dz = -\int_{C_4} E(z)\, dz,$$

so that $r = 0$. When one or more poles of $E(z)$ lie on the boundary L of $\mathfrak{P}$, then it is always possible to effect a small translation which takes L to a new curve L^* such that all the poles of $E(z)$ in $\mathfrak{P}$ are in the interior of L^*, and no poles of $E(z)$ are on L^* (Figure 32); but then, as just shown, $r = 0$.

If an elliptic function $E(z)$ has no poles in the period parallelogram, then it is regular on its closure and, being continuous, it must assume its maximum at some point z_1 of the closure of $\mathfrak{P}$. But then, in view of the periodicity of $E(z)$, z_1 is a maximum of $E(z)$ for arbitrary z and so $E(z)$ is constant. It follows that a nonconstant elliptic function has at least one pole in $\mathfrak{P}$. In view of Theorem 1, $E(z)$ cannot have one simple pole in $\mathfrak{P}$ and, therefore, $E(z)$ has at least two simple poles or one double pole in $\mathfrak{P}$. The $\wp$-function is in a sense the simplest example of a nonconstant elliptic function, for it is

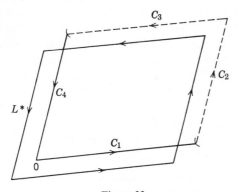

Figure 32

an elliptic function with just one double pole (at $z = 0$ with principal part z^{-2}) in the period parallelogram.

Theorem 2: The number of zeros of a nonconstant elliptic function on $\mathfrak{P}$ is equal to the number of poles, provided that each zero and each pole is counted in accordance with its multiplicity.

Proof: The derivative $E'(z)$ of an elliptic function is likewise an elliptic function, for $E'(z)$ is meromorphic and has all the ω as periods. Assume that $E(z)$ is not constant. Then $E'(z)/E(z)$ is an elliptic function. By Theorem 1, the sum of the residues of $E'(z)/E(z)$ is 0. On the other hand, if $E(z)$ has neither zeros nor poles on the boundary L of $\mathfrak{P}$, then the sum of the residues of the function $E'(z)/E(z)$ is equal to the difference between the number of zeros and the number of poles of $E(z)$. It follows that the two numbers are the same. Should there be zeros or poles of $E(z)$ on L, we can change matters in our favor by means of a small translation of L as was done in the proof of Theorem 1.

The number of poles of an elliptic function in $\mathfrak{P}$, each counted according to its multiplicity, is called its *order*.

Theorem 3: Counting multiplicities, a nonconstant elliptic function of order h takes on every complex value h times in $\mathfrak{P}$.

Proof: Let $E(z)$ be a nonconstant elliptic function of order h and c an arbitrary constant. The function $F(z) = E(z) - c$ has the same poles with the same principal parts as the function $E(z)$, so that $F(z)$ is also a non-constant elliptic function whose order for every c is h. The zeros of $F(z)$ coincide in value and multiplicity with the solutions of the equation $E(z) = c$. Application of Theorem 2 to the function $F(z)$ completes the proof.

Since the function $\wp(z)$ has exactly one double pole at $z = 0$ in the period parallelogram, it has order $h = 2$. This fact and Theorem 3 show that $\wp(z)$ takes on in the period parallelogram every preassigned value exactly twice, either simply at two distinct points or doubly at a single point. This consequence of Theorem 3 was used in the previous section. We note that already in Section 11 it was shown that the function f of that section assumes in $\mathfrak{P}$ every value exactly twice and, since $\wp(z)$ is a fractional linear function of f, the same is true of $\wp(z)$. At the same time we wish to point out that the inversion problem was not solved in Section 11.

By the *sum of the zeros of a nonconstant elliptic function of order h* we mean the sum $a_1 + \cdots + a_h$ of its zeros $a_1, \ldots, a_h$ in $\mathfrak{P}$ each appearing as many times as its multiplicity indicates; a similar definition applies to the sum $b_1 + \cdots + b_h$ of its poles $b_1, \ldots, b_h$ in $\mathfrak{P}$.

Theorem 4: The sum of the zeros of a nonconstant elliptic function differs from the sum of its poles by a period.

Proof: Let $g(z)$ be a function which is regular on $\mathfrak{P}$ and on its boundary L. With the given elliptic function $E(z)$ we construct the function

$$G(z) = g(z) \frac{E'(z)}{E(z)}.$$

Apart from possible poles this function is regular on $\mathfrak{P}$ and L; specifically, its possible poles are found among the zeros and poles of the denominator $E(z)$. As before, let $a_1, \ldots, a_h$ and $b_1, \ldots, b_h$ be the zeros and poles of $E(z)$ in $\mathfrak{P}$. If $a_1 = \cdots = a_k = a$ is a zero of order k, then there corresponds to it the Laurent expansion

$$\frac{E'(z)}{E(z)} = \frac{k}{z-a} + \cdots = \sum_{n=1}^{k} \frac{1}{z-a_n} + \cdots,$$

and if $b_1 = \cdots = b_l = b$ is a pole of order l, then there corresponds to it the Laurent expansion

$$\frac{E'(z)}{E(z)} = -\frac{l}{z-b} + \cdots = -\sum_{n=1}^{l} \frac{1}{z-b_n} + \cdots.$$

Now assume to begin with that there are no zeros or poles of $E(z)$ on L. Then we have by the residue theorem

$$\frac{1}{2\pi i} \int_{L} G(z)\, dz = \sum_{n=1}^{h} g(a_n) - \sum_{n=1}^{h} g(b_n).$$

In particular, for $g(z) = z$ the expression on the right-hand side is precisely the difference between the zeros and poles of the function $E(z)$. We must now show that the integral

$$\Omega = \frac{1}{2\pi i} \int_{L} \frac{zE'(z)}{E(z)}\, dz$$

on the left is a period of the elliptic function $E(z)$. Note that the sides C_3 and C_2 of the parallelogram arise from the sides C_1 and C_4 when we replace z by $z + \omega_2$ and $z + \omega_1$ and reverse directions. Therefore,

$$2\pi i\Omega = \int_{C_1} [z - (z+\omega_2)] \frac{E'(z)}{E(z)}\, dz + \int_{C_4} [z - (z+\omega_1)] \frac{E'(z)}{E(z)}\, dz$$

$$= \omega_1 \int_{0}^{\omega_2} \frac{E'(z)}{E(z)}\, dz - \omega_2 \int_{0}^{\omega_1} \frac{E'(z)}{E(z)}\, dz$$

$$= \omega_1 [\log E(z)]_0^{\omega_2} - \omega_2 [\log E(z)]_0^{\omega_1}.$$

The expressions $[\log E(z)]_0^{\omega_1}$ and $[\log E(z)]_0^{\omega_2}$ represent the change in an arbitrary but fixed branch of the multiple-valued function $\log E(z)$ on the segments from 0 to ω_1 and from 0 to ω_2. The periodicity of $E(z)$ shows that these changes are multiples of $2\pi i$, say, $2\pi i n_1$ and $2\pi i n_2$ with integer n_1 and n_2. It follows that

$$\Omega = n_2\omega_1 - n_1\omega_2$$

is indeed a period. If L is not free of poles and zeros of $E(z)$, then we obtain the same result by first subjecting L to a suitable small translation.

We shall now show that the conditions which are consequences of Theorems 2 and 4 guarantee the existence of an elliptic function with preassigned zeros and poles in the period parallelogram. Specifically, we have

Theorem 5: Let a_k, b_l ($k, l = 1, \ldots, h$) be $2h$ given numbers in $\mathfrak{P}$ such that $a_k \neq b_l$. If the two sums $a_1 + \cdots + a_h$ and $b_1 + \cdots + b_h$ differ by a period, then there exists an elliptic function whose zeros in $\mathfrak{P}$ are precisely the a_k and whose poles in $\mathfrak{P}$ are precisely the b_l, and all such functions differ from it by a nonzero multiplicative constant.

Proof: If $E_1(z)$ and $E_2(z)$ are two elliptic functions which have in $\mathfrak{P}$ precisely the prescribed zeros and poles, then their ratio $E_1(z)/E_2(z)$ is an elliptic function of order zero with no zeros, that is, a nonzero constant. It remains to prove the existence of a function $E(z)$ of the required kind.

In the trivial case $h = 0$ we can take $E(z) = 1$. The case $h = 1$ is ruled out, since for two distinct points a_1, b_1 in $\mathfrak{P}$ the difference $a_1 - b_1$ cannot be equal to a period. We may therefore suppose $h \geq 2$. The general case can easily be reduced to the special case $b_1 = \cdots = b_h = 0$, and it is this special case which we are about to investigate. By assumption in this special case the sum

$$(1) \qquad\qquad\qquad a_1 + \cdots + a_h = \omega$$

is a period. For $r = 1, 2, \ldots$ we form the elliptic functions

$$p_{2r} = [\wp(z)]^r, \qquad p_{2r+1} = \wp'(z)[\wp(z)]^{r-1}$$

and put $p_1 = 1$. The function p_n ($n = 2, 3, \ldots$) has precisely one pole on $\mathfrak{P}$, namely, the pole of order n at $z = 0$; p_1, of course, has no poles. We propose to find h constants $c_1, \ldots, c_h$ such that the elliptic function

$$E(z) = c_1p_1 + \cdots + c_hp_h$$

has in $\mathfrak{P}$ precisely the prescribed zeros $a_1, \ldots, a_h$. Since $E(z)$ is regular in $\mathfrak{P}$ for $z \neq 0$, Theorem 2 implies that $E(z)$ has at the origin a pole of order h, so that $c_h \neq 0$.

Let $\alpha_1, \ldots, \alpha_q$ be the different numbers in the sequence $a_1, \ldots, a_h$, and let n_l ($l = 1, \ldots, q$) denote the multiplicity of α_l so that

$$n_1 + \cdots + n_q = h.$$

We must choose the numbers $c_1, \ldots, c_h$ so as to satisfy the h equations

$$E^{(k)}(\alpha_l) = 0 \qquad (k = 0, \ldots, n_l - 1; \ l = 1, \ldots, q).$$

With

$$E^{(k)}(z) = c_1 p_1^{(k)} + \cdots + c_h p_h^{(k)}$$

we obtain h homogeneous linear equations for the h unknowns $c_1, \ldots, c_h$. If the determinant of this system of linear equations were $\neq 0$, then the system would have only the trivial solution $c_1 = \cdots = c_h = 0$, which is useless for our purpose. Instead of proving that the determinant in question vanishes we disregard the last equation

(2) $$E^{(n_1-1)}(\alpha_q) = 0,$$

which we suppose to correspond to the zero a_h, and prove that a definite nontrivial solution $c_1, \ldots, c_h$ of the remaining $h - 1$ equations must also satisfy the disregarded equation. Since the $h - 1$ functions $p_2, \ldots, p_h$ have poles of different orders at $z = 0$ and p_1 is finite at $z = 0$, it follows that the constructed function $E(z)$ does not vanish identically, and, on the other hand, has at least the $h - 1$ zeros $a_1, \ldots, a_{h-1}$. But $E(z)$ can have a pole in $\mathfrak{P}$ only at $z = 0$ and the order of the pole is at most h, so that, by Theorem 2, the order j of $E(z)$ is $h - 1$ or h. The sum of the poles of $E(z)$ is, in any case, equal to 0. If $j = h - 1$, then $a_1, \ldots, a_{h-1}$ would be all the zeros of $E(z)$ in $\mathfrak{P}$ and, by Theorem 4, the sum of the zeros

$$a_1 + \cdots + a_{h-1} = \omega^*$$

would be a period. But then (1) shows that $\omega - \omega^* = a_h$ would be a period when a_h is a nonzero point of $\mathfrak{P}$. It follows that $j = h$ and the function $E(z)$ has in addition to $a_1, \ldots, a_{h-1}$ exactly one more zero a in $\mathfrak{P}$. According to Theorem 4 the sum of the zeros $a_1 + \cdots + a_{h-1} + a$ must be a period and the same must be true of the difference $a - a_h$, which implies that $a = a_h$. We conclude that the disregarded equation (2) is also satisfied. This proves our theorem in the special case $b_1 = \cdots = b_h = 0$. The analogous special case $a_1 = \cdots = a_h = 0$ can be reduced to the special case just proved by interchanging the a_k with the b_l and replacing the corresponding elliptic function by its reciprocal.

It remains to consider the general case. We may assume that not all the a_k and not all the b_l vanish. We choose the point a_0 in $\mathfrak{P}$ so that the sum $a_0 + a_1 + \cdots + a_h$ is a period. If we put $b_0 = a_0$ then, by assumption, the corresponding sum $b_0 + b_1 + \cdots + b_h$ is also a period. Suppose that exactly m of the numbers $a_0, \ldots, a_h$ are 0 and that exactly n of the numbers $b_0, \ldots, b_h$ are zero; m and n are both less than $h + 1$. If we disregard the numbers a_k $(k = 0, \ldots, h)$ which are equal to 0, then the sum of the remaining $h + 1 - m$ numbers is a period and, in view of what has already been proved, we can assert the possibility of finding constants $c_1, \ldots, c_\mu$ such that the elliptic function

(3)
$$E_1(z) = c_1 p_1 + \cdots + c_\mu p_\mu$$

has in $\mathfrak{P}$ precisely the zeros $a_k \neq 0$, where the order $\mu = h + 1 - m > 1$ and $c_\mu \neq 0$. Similarly, it is possible to find constants $d_1, \ldots, d_\nu$ such that the elliptic function

(4)
$$E_2(z) = d_1 p_1 + \cdots + d_\nu p_\nu$$

has in $\mathfrak{P}$ precisely the zeros $b_l \neq 0$, where the order $\nu = h + 1 - n > 1$ and $d_\nu \neq 0$. Now we form the quotient

$$E(z) = \frac{E_1(z)}{E_2(z)}.$$

If $a_0 \neq 0$, then $a_0 = b_0$ is a common zero of $E_1(z)$ and $E_2(z)$. In any case, the $a_k \neq 0$ with $k > 0$ appear as zeros of $E_1(z)$ and the $b_l \neq 0$ with $l > 0$ appear as zeros of $E_2(z)$. It remains to investigate the behavior of $E(z)$ at the origin. Since $\mu - \nu = n - m$, it follows that $E(z)$ has at $z = 0$ a zero of order $m - n$, a pole of order $n - m$, or a value other than 0 and ∞, according as $m > n, m < n$, or $m = n$. But this is precisely what we are supposed to prove.

Theorem 6:   Every elliptic function $E(z)$ has a unique representation

$$E(z) = S(x) + yT(x), \qquad x = \wp(z), \quad y = \wp'(z),$$

where $S(x)$ and $T(x)$ are rational functions of x with constant coefficients, and, conversely, every such expression $S(\wp) + \wp'T(\wp)$ is an elliptic function.

Proof: Let h be the order of the elliptic function $E(z)$. If $h = 0$ then $E(z)$ is a constant c, and we can take $S(x) = c$, $T(x) = 0$. Now let $h > 0$ and let $a_1, \ldots, a_h$ be the zeros and $b_1, \ldots, b_h$ the poles of $E(z)$ in $\mathfrak{P}$. In view of Theorem 4 the difference between the sum $a_1 + \cdots + a_h$ of the zeros of $E(z)$ and the sum $b_1 + \cdots + b_h$ of its poles is a period. As was shown in the proof of Theorem 5 it is possible to construct two elliptic functions $E_1(z)$ and $E_2(z)$ of the form given in (3) and (4) whose quotient $E_1(z)/E_2(z)$ differs from $E(z)$ only by a nonzero multiplicative constant. Including this constant in $E_1(z)$ we get

(5)
$$E(z) = \frac{E_1(z)}{E_2(z)}.$$

With $p_1, p_2, \ldots$ as defined and with $\wp(z) = x$, $\wp'(z) = y$ we have

$$E_1(z) = A(x) + yB(x), \qquad E_2(z) = C(x) + yD(x),$$

where $A(x), B(x), C(x), D(x)$ are four polynomials in x with constant coefficients. Now, $E_2(z)$ is not identically zero so that the two polynomials $C(x)$

and $D(x)$ do not both vanish identically in x. Since $\wp'(z)$ is an odd function of z and $\wp(z)$ is an even function of z, it follows that the function

$$E_0(z) = C(x) - yD(x)$$

is not identically zero. Because of the differential equation for $\wp(z)$ we have

$$y^2 = 4x^3 - g_2 x - g_3,$$
$$E_0(z)E_1(z) = A_1(x) + yB_1(x), \qquad E_0(z)E_2(z) = C_1(x)$$

with certain polynomials $A_1(x)$, $B_1(x)$, $C_1(x)$, of which $C_1(x)$ does not vanish identically in x. Putting $S(x) = A_1(x)/C_1(x)$, $T(x) = B_1(x)/C_1(x)$, we obtain the required formula.

If

$$E(z) = S^*(x) + yT^*(x), \qquad x = \wp(z), \quad y = \wp'(z),$$

with rational functions $S^*(x)$, $T^*(x)$, then putting $S(x) - S^*(x) = S_0(x)$, $T(x) - T^*(x) = T_0(x)$, we obtain

$$S_0(x) + yT_0(x) = 0.$$

Since $x = \wp(z)$ is even and $y = \wp'(z)$ is odd, we conclude that $S_0(x) = 0$, $T_0(x) = 0$ for all x. This proves the uniqueness of the representation $E(z) = S(x) + yT(x)$, for $\wp(z)$ itself satisfies no polynomial equation with complex coefficients.

Since the elliptic functions form a field it follows that, conversely, for every choice of rational functions $S(x)$, $T(x)$ the expression $S(x) + yT(x)$ with $x = \wp(z)$ and $y = \wp'(z)$ yields an elliptic function. This completes the proof of the theorem.

Theorem 6 includes an earlier assertion to the effect that all the elliptic functions are expressible as rational functions of $\wp$ and $\wp'$. Further, if

$$E_1(z) = S_1(\wp) + \wp'T_1(\wp), \qquad E_2(z) = S_2(\wp) + \wp'T_2(\wp)$$

are two elliptic functions, then it is possible to eliminate $\wp$ and $\wp' = \sqrt{4\wp^3 - g_2\wp - g_3}$ and obtain a polynomial equation in $E_1(z)$ and $E_2(z)$ with constant coefficients. It follows that any two elliptic functions are algebraically dependent.

We shall now present a different formulation of the results obtained so far. To this end we map the parallelogram $\mathfrak{P}$ in the z-plane conformally onto the surface $\mathfrak{R}_0$ lying over the complex x-sphere using the relations $x = \wp(z)$, $y = \wp'(z)$; here $\mathfrak{R}_0$ is obtained from the Riemann surface $\mathfrak{R}$ of the function

$$(6) \qquad y = \sqrt{4x^3 - g_2 x - g_3}$$

by means of the canonical dissection corresponding to the sides of $\mathfrak{P}$. Hence x is the new independent variable and if we associate with it the value of the square root y, then we obtain in this way precisely one point of $\mathfrak{R}$. This is also true for the branch points e_1, e_2, e_3, or ∞, for y takes on there the uniquely determined value 0 or ∞, and at these points conformality of the mapping refers, as before, to the local uniformizing parameter. Now we consider a function element $f(x)$ which can be uniquely continued analytically to all the points of $\mathfrak{R}$. (Here we recall that poles are admitted and one is always to use the local parameter.) A function given in this manner is said to be meromorphic on $\mathfrak{R}$. This definition makes it clear that the meromorphic functions on $\mathfrak{R}$ form a field. Since x and y are meromorphic on $\mathfrak{R}$, so is, more generally, every rational function of x and y whose denominator, considering (6), does not vanish identically in x. In view of (6) such a rational function can again be given the form $S(x) + yT(x)$ with rational $S(x)$ and $T(x)$. We now prove the converse of this last result.

Theorem 7: Every function meromorphic on $\mathfrak{R}$ is of the form $S(x) + yT(x)$ with rational $S(x)$ and $T(x)$.

Proof: Let $f(x)$ be meromorphic on $\mathfrak{R}$. By means of the conformal mapping $x = \wp(z)$ we obtain on the period parallelogram $\mathfrak{P}$ the function

$$g(z) = f[\wp(z)],$$

which is regular apart from poles. Since $\wp(z)$ has the periods ω_1, ω_2 and $f(x)$ is single-valued on $\mathfrak{R}$, it follows that $g(z)$ also has these periods and is therefore an elliptic function of the variable z. Now Theorem 6 implies the asserted result

Since x and y are connected by the same algebraic relations as $\wp(z)$ and $\wp'(z)$, it follows, in view of Theorems 6 and 7, that the correspondence $x = \wp(z)$, $y = \wp'(z)$ establishes an isomorphism between the field of meromorphic functions on $\mathfrak{R}$ and the field of elliptic functions. Theorems 2 and 3 can now be formulated so as to apply to the meromorphic functions on $\mathfrak{R}$. The order h of $f(x)$ is to stand for the number of poles on $\mathfrak{R}$. If the function is not a constant, then it assumes on $\mathfrak{R}$ every value exactly h times. We also consider, for comparison, the functions on the complex number sphere $\mathfrak{K}$ which are single-valued and regular apart from poles. These functions coincide with the rational functions in the variable x; this well-known fact follows readily from Theorem 7. For these functions we define the order h to be the number of poles on the simple number sphere $\mathfrak{K}$. Unless it is a constant, such a function assumes on $\mathfrak{K}$ every value exactly h times. It is possible to prescribe the h zeros and the h poles of a rational function of order h arbitrarily. In fact, every such function with zeros $a_1, \ldots, a_h$ and poles $b_1, \ldots, b_h$ is given by

$$f(x) = c \, \frac{(x - a_1) \cdots (x - a_h)}{(x - b_1) \cdots (x - b_h)}$$

with constant $c \neq 0$, where we delete a factor corresponding to a possible zero or pole ∞. This property does not carry over to the meromorphic functions on $\Re$, for the h zeros and h poles must satisfy a condition which can be deduced from Theorems 4 and 5. In fact, let $\mathfrak{a}_1, \ldots, \mathfrak{a}_h$ and $\mathfrak{b}_1, \ldots, \mathfrak{b}_h$ be the prescribed zeros and poles on $\Re$ of the required function $f(x)$. The substitution $x = \wp(z)$, $y = \wp'(z)$ carries these zeros and poles into the zeros $a_1, \ldots, a_h$ and poles $b_1, \ldots, b_h$ of the elliptic function $f(\wp(z))$ on $\mathfrak{P}$, and these are subject to the restriction that the sum $a_1 + \cdots + a_h$ differs from the sum $b_1 + \cdots + b_h$ by a period. The mapping of $\Re$ onto $\mathfrak{P}$ is effected by means of the elliptic integral

$$z = \int_\infty^x \frac{dx}{y}$$

where the path of integration on the canonically dissected Riemann surface $\Re_0$ extends from the branch point ∞ to the point $\mathfrak{x}$ over x determined by the value of y. Hence the required condition is

$$\sum_{k=1}^h \int_\infty^{\mathfrak{a}_k} \frac{dx}{y} - \sum_{k=1}^h \int_\infty^{\mathfrak{b}_k} \frac{dx}{y} = \omega,$$

where ω is a suitable period, and where, for the sake of uniqueness, the upper limits of the integrals are the appropriate points on the Riemann surface. Here the paths of integration can run arbitrarily on the uncut surface $\Re$, for the necessary modification can be combined with ω. The condition can be given the simpler form

$$\sum_{k=1}^h \int_{\mathfrak{b}_k}^{\mathfrak{a}_k} \frac{dx}{y} = \omega,$$

and is necessary and sufficient for the existence on $\Re$ of an analytic function of order h with zeros $\mathfrak{a}_1, \ldots, \mathfrak{a}_h$ and poles $\mathfrak{b}_1, \ldots, \mathfrak{b}_h$. Moreover, a set of zeros and poles satisfying the existence condition determines the function to within a nonzero multiplicative constant. This result is a special case of the so-called Theorem of Abel which we will deduce much later in Chapter 4 (Vol. II) in connection with the investigation of abelian integrals.

We close this section with an investigation of the *general elliptic integral*

$$w = \int_{x_0}^x R(x, y)\, dx \qquad (y = \sqrt{4x^3 - g_2 x - g_3},\ g_2^3 - 27g_3^2 \neq 0).$$

Here the integrand is an arbitrary function of the independent variable x which is analytic on $\Re$ and which we write as a rational function of x and y; the integration extends over a prescribed path between the points $\mathfrak{x}_0$ and $\mathfrak{x}$ of $\Re$. Suppose the integral to be regular at $\mathfrak{x}_0$. If $\mathfrak{x}_0$ is not a branch point and lies over x_0, then, with rectilinear integration, w is given in a neighborhood of $\mathfrak{x}_0$ by a power series in $(x - x_0)$. In particular, if $R(x, y) = c/y$ with constant

c, then w differs from the elliptic integral of the first kind in the Weierstrass normal form only by the factor c and by an additive constant. It follows that, in this case, the function element of w given at x_0 can be continued analytically on $\Re$ and is finite because of the absence of poles. We now show that, conversely, if w can be continued analytically along every path on $\Re$ and the function has no poles, then we must have $R(x, y) = c/y$. In fact, let us again put $x = \wp(z)$, $y = \wp'(z)$, where the point $z = z_0$ of $\mathfrak{P}$ corresponds to the beginning point x_0. Then $w = w(z)$ is a function of z which is regular in a neighborhood of z_0 and can be continued analytically along every path in the z-plane issuing from z_0 without encountering poles. By the monodromy theorem $w(z)$ is single-valued everywhere and therefore an entire function of z. In the vicinity of z_0 we have

$$\frac{dw}{dz} = \frac{dw}{dx}\frac{dx}{dz} = R(x, y)y.$$

The validity of this relation on all of $\mathfrak{P}$ follows by analytic continuation. Since the derivative dw/dz is regular on $\mathfrak{P}$, the meromorphic function $yR(x, y)$ on $\Re$ has no poles. As such it is a constant c, and we have $R(x, y) = c/y$, $dw/dz = c$. This completes the proof.

15. The addition theorem

Let $f(\xi, \eta, \zeta)$ be an irreducible homogeneous polynomial in three variables ξ, η, ζ of degree $n > 0$ with complex coefficients; thus f cannot be written as a product of two polynomials of lower degree. If we interpret ξ, η, ζ as homogeneous coordinates in the projective plane, where we allow complex values, then the totality of nontrivial solutions of the equation $f(\xi, \eta, \zeta) = 0$ forms an irreducible algebraic curve of degree n or, briefly, a curve of degree n. A double point of this curve is a point whose coordinates satisfy the three equations $f_\xi = 0$, $f_\eta = 0$, $f_\zeta = 0$. In view of Euler's theorem we have the identity $\xi f_\xi + \eta f_\eta + \zeta f_\zeta = nf$, so that every double point actually lies on the curve. For an algebraic curve to have one or more double points the coefficients of its equation must satisfy certain conditions which can be obtained by elimination. The double points can also be characterized by the property that every line through a double point intersects the n-th degree curve in at most $n - 2$ additional points, whereas, in general, a line through an arbitrary point on the curve has $n - 1$ additional points in common with the curve. Linear transformations of the coordinates ξ, η, ζ preserve the degree of the curve, and a double point remains a double point in the new coordinates. We shall be concerned with cubic curves free of double points. It is shown in algebra and in geometry that there exists a suitable projective transformation which reduces the curve to the normal form

$$4\xi^3 - g_2\xi\zeta^2 - g_3\zeta^3 - \eta^2\zeta = 0$$

with certain constants g_2 and g_3. If we put the three partial derivatives $12\xi^2 - g_2\zeta^2$, $-2\eta\zeta$, $-2g_2\xi\zeta - 3g_3\zeta^2 - \eta^2$ equal to 0, then $\zeta = 0$ implies $\xi = 0$, $\eta = 0$, which does not define a point of the curve. $\zeta \neq 0$ implies $\eta = 0$, $g_2 = 12\xi^2\zeta^{-2}$, $3g_3 = -2g_2\xi\zeta^{-1}$; that is, $27g_3^2 = g_2^3$. In other words, the condition $g_2^3 - 27g_3^2 \neq 0$ states that the cubic curve in normal form has no double points. If we introduce nonhomogeneous coordinates then we obtain the normal form

$$(1) \qquad y^2 = 4x^3 - g_2 x - g_3 \qquad (g_2^3 - 27g_3^2 \neq 0).$$

In this way the Riemann surface $\Re$ of the function $y = \sqrt{4x^3 - g_2 x - g_3}$ is mapped in a one-one manner onto the cubic curve without double points in the projective plane with homogeneous coordinates ξ, η, ζ; in fact, $\Re$ is the result of projection of the curve onto the x-axis, where x is a complex variable. If we introduce the corresponding $\wp$-function and put $x = \wp(z)$, $y = \wp'(z)$, then we obtain a parametric representation of the curve; at the same time the period parallelogram $\mathfrak{P}$ is mapped in a one-one manner onto the curve. Finally, we can invert the projective transformation which produced the normal form and obtain thereby a parametric representation in terms of elliptic functions for the nonhomogeneous coordinates x, y of an arbitrary cubic without double points.

We shall now deduce the addition therem for the $\wp$-function by means of a simple geometric argument. Our starting point is a cubic curve in the normal form (1). When we introduce homogeneous coordinates then every line intersects the curve in three points of which two, or even all three, may coincide, as in the case of a tangent line or a line through a point of inflection. Determination of the coordinates of the three points of intersection leads to a cubic equation. If the coordinates of two points of intersection are known, then, as we are about to verify, the coordinates of the third point of intersection can be expressed rationally in terms of the known coordinates. In fact, let x_k, y_k $(k = 1, 2, 3)$ be the nonhomogeneous coordinates of the three points of intersection $\mathfrak{y}_k$, where we suppose x_1 and x_2 distinct and finite. If

$$y = ax + b$$

is the equation of the secant through the points $\mathfrak{y}_1$ and $\mathfrak{y}_2$, then

$$a = \frac{y_1 - y_2}{x_1 - x_2}.$$

In view of (1) the cubic equation

$$4x^3 - (ax + b)^2 - g_2 x - g_3 = 0$$

has exactly the three solutions x_1, x_2, x_3. Since the coefficient of x^3 in our equation is 4, the sum $x_1 + x_2 + x_3$ of the roots is the product of $-1/4$ and the coefficient of x^2, that is

$$x_1 + x_2 + x_3 = \frac{a^2}{4},$$

so that

(2) $$x_3 = \frac{1}{4}\left(\frac{y_1 - y_2}{x_1 - x_2}\right)^2 - x_1 - x_2.$$

The corresponding formula for y_3 follows from

(3) $$\begin{vmatrix} x_1 & x_2 & x_3 \\ y_1 & y_2 & y_3 \\ 1 & 1 & 1 \end{vmatrix} = 0.$$

Suppose that under the parametric representation $x = \wp(z)$, $y = \wp'(z)$ there correspond to the three points of intersection $\mathfrak{y}_k$ with coordinates x_k, y_k the values z_k on $\mathfrak{P}$. Since the elliptic function

$$y - ax - b = \wp'(z) - a\wp(z) - b$$

has exactly one pole on $\mathfrak{P}$, namely the triple pole at $z = 0$, it follows from Theorem 4 that it must have exactly three zeros and these are just the values z_1, z_2, z_3, where z_3 may coincide with z_1 or z_2. Further, the sum of the poles is 0; so that, in view of Theorem 4, the sum of the roots $z_1 + z_2 + z_3 = \omega$ is a period. Since $\wp(z)$ is an even function it follows that

$$x_3 = \wp(\omega - z_1 - z_2) = \wp(-z_1 - z_2) = \wp(z_1 + z_2),$$

and (2) yields

(4) $$\wp(z_1 + z_2) = \frac{1}{4}\left(\frac{\wp'(z_1) - \wp'(z_2)}{\wp(z_1) - \wp(z_2)}\right)^2 - \wp(z_1) - \wp(z_2).$$

This is the addition theorem for the $\wp$-function. In this derivation x_1 and x_2 were supposed distinct and finite. To fulfill this requirement we take as z_1 an arbitrary point in the interior of $\mathfrak{P}$ and as $z_2 \neq z_1$ an arbitrary point in a sufficiently small neighborhood of z_1. Since $\wp(z)$ and $\wp'(z)$ are meromorphic functions we conclude that the addition theorem holds generally for independent variables z_1 and z_2.

In this derivation of the addition theorem we made use of the fact that the coordinates of the third point of intersection of the secant with the cubic curve can be expressed rationally through the coordinates of the other two points of intersection. This fact was first used by Fermat in his investigations of cubic diophantine equations; the so-called Fermat method of secants has played a very important role in the theory of numbers.

Hindsight makes it possible to verify the addition theorem by direct computation. We choose a fixed value for z_2 such that $\wp(z_2)$ is finite and $\wp'(z_2) \neq 0$ and keep $z_1 = z$ variable. Then the difference of the two sides of (4) is an elliptic function $\varphi(z)$. The possible poles of $\varphi(z)$ differ from $z = 0$ and $z = -z_2$ by a period. In the vicinity of $z = 0$ we have

$$\wp(z) = z^{-2} + b_1 z^2 + \cdots, \qquad \wp'(z) = -2z^{-3} + 2b_1 z + \cdots,$$

$$\frac{1}{4}\left[\frac{\wp'(z) - \wp'(z_2)}{\wp(z) - \wp(z_2)}\right]^2 - \wp(z) = z^{-2}\left[\frac{1 - \frac{1}{2}\wp'(z_2)z^3 + \cdots}{1 - \wp(z_2)z^2 + \cdots}\right]^2 - z^{-2} - b_1 z^2 + \cdots$$

$$= 2\wp(z_2) + \cdots,$$

so that $\varphi(z)$ is regular at $z = 0$ and $\varphi(0) = 0$. Further, in the vicinity of $z = -z_2$ we have

$$\wp(z + z_2) = (z + z_2)^{-2} + \cdots,$$

$$\frac{1}{4}\left[\frac{\wp'(z) - \wp'(z_2)}{\wp(z) - \wp(z_2)}\right]^2 = \left[\frac{\wp'(z_2) - \frac{1}{2}\wp''(z)(z + z_2) + \cdots}{\wp'(z_2)(z + z_2) - \frac{1}{2}\wp''(z_2)(z + z_2)^2 + \cdots}\right]^2$$

$$= (z + z_2)^{-2} + \cdots,$$

so that $\varphi(z)$ is also regular at $z = -z_2$. As an entire elliptic function $\varphi(z)$ is a constant, namely the constant 0.

If in the addition theorem we make use of the differential equation $\wp' = \sqrt{4\wp^3 - g_2\wp - g_3}$ and eliminate the square roots by squaring, then we obtain from (4) an algebraic equation with constant coefficients connecting the three functions $\wp(z_1 + z_2)$, $\wp(z_1)$, $\wp(z_2)$. This result can be extended to all elliptic functions $E(z)$. In fact, by Theorem 6 of the preceding section we have

$$E(z) = S(\wp(z)) + \wp'(z)T(\wp(z)),$$

with rational functions $S(x)$ and $T(x)$. This implies the existence of an algebraic equation

$$H[E(z), \wp(z)] = 0,$$

where $H(u, v)$ is an irreducible polynomial in u and v with constant coefficients. We might mention that $H(u, v)$ can be chosen so as to be of degree one or two in u. Elimination of the three $\wp$-functions from the three equations

$$H[E(z_1), \wp(z_1)] = 0, \qquad H[E(z_2), \wp(z_2)] = 0,$$

$$H[E(z_1 + z_2), \wp(z_1 + z_2)] = 0$$

and the algebraic equation connecting $\wp(z_1 + z_2)$, $\wp(z_1)$ and $\wp(z_2)$ yields a relation of the form

$$A(E(z_1 + z_2), E(z_1), E(z_2)) = 0,$$

where $A(u, v, w)$ is an irreducible polynomial with constant coefficients. This relation states that with every elliptic function there is associated an algebraic addition theorem. Weierstrass established the converse result that a meromorphic function has an algebraic addition theorem only if it is an elliptic function or a degenerate elliptic function; degenerate elliptic functions will be discussed in the next section.

We shall now formulate the addition theorem for the $\wp$-function in terms of the corresponding property of elliptic integrals of the first kind. For the three points of intersection $x_k = \wp(z_k)$, $y_k = \wp'(z_k)$ ($k = 1, 2, 3$) of the curve $4x^3 - g_2 x - g_3 - y^2 = 0$ with a line we established the equation

$$z_1 + z_2 = \omega - z_3,$$

where ω is a suitably chosen period. On the other hand, we have

$$z_k = \int_\infty^{x_k} \frac{dx}{y} \qquad (k = 1, 2, 3),$$

where we integrate on the canonically dissected Riemann surface $\mathfrak{R}_0$. The end point of each of the three paths of integration is uniquely determined on $\mathfrak{R}$ by prescribing the pair x_k, y_k. We now replace the path of integration in the integral z_3 by its mirror image obtained by interchanging the two sheets of $\mathfrak{R}$. Then y must be replaced by $-y$, and for the new path of integration we have

$$-z_3 = \int_\infty^{x_3} \frac{dx}{y}.$$

By adjoining a suitable closed path of integration we can include the period ω in the value of the last integral and obtain the relation

$$(5) \qquad \int_\infty^{x_1} \frac{dx}{y} + \int_\infty^{x_2} \frac{dx}{y} = \int_\infty^{x_3} \frac{dx}{y},$$

after a suitable choice of the third path of integration on the Riemann surface $\mathfrak{R}$. The upper limit x_3 as well as the ordinate y_3 are definite rational functions of x_1, y_1 and x_2, y_2 specified by the formulas (2) and (3). This, then, is the required addition theorem for elliptic integrals of the first kind.

In the second section we gave a different proof of the Euler addition theorem and it is clearly desirable to establish a link between the formulas (5) and (6) in Section 2 and the formulas (2) and (5) of the present section. This requires the use of a relation connecting the forms of Legendre and Weierstrass which we do not wish to investigate in all generality. In the case of the lemniscatic integral this relation is readily obtained by making the change of variable $u = x^{-1/2}$. We then have

$$\int_0^u \frac{du}{\sqrt{1 - u^4}} = -\int_\infty^x \frac{dx}{\sqrt{4x^3 - 4x}},$$

which gives the special case $g_2 = 4$, $g_3 = 0$ of the Weierstrass normal form. For this case it is not very difficult to transform the formula (2) of the present section into the formula (3) of Section 2. We leave this task to the reader.

16. Degenerate elliptic functions

We conclude this chapter with a discussion of *degenerate elliptic functions* which arise when one or both basic periods are allowed to become infinite. Since the trigonometric functions and the exponential functions possess addition theorems and are, at any rate, simply periodic, it is natural to expect that these functions are degenerate elliptic functions. So far, when investigating elliptic functions in general and the $\wp$-function in particular, we kept the pair of basic periods ω_1, ω_2 fixed. In the definition of the $\wp$-function the essential thing is the period lattice and not a particular one of its bases. We can therefore make the assumption that the pair ω_1, ω_2 is reduced in the sense of Section 9. This means that the quotient $\omega_2/\omega_1 = \tau = \xi + \eta i$ lies in the region $\mathfrak{B}$ of the upper half plane defined by the inequalities

$$\xi^2 + \eta^2 \geq 1, \qquad -\tfrac{1}{2} \leq \xi \leq \tfrac{1}{2}, \qquad \eta > 0.$$

Since we shall no longer regard the period lattice as fixed we adopt the more accurate notation

$$\wp(z) = \wp(z; \ \omega_1, \ \omega_2)$$

according to which the $\wp$-function depends on three complex variables. We now let the pair ω_1, ω_2 vary over an infinite sequence $\omega_1 = \omega_{1k}$, $\omega_2 = \omega_{2k}$ $(k = 1, 2, \ldots)$, with $\omega_{1k} \neq 0$ and $\omega_{2k}/\omega_{1k} = \tau_k$ in $\mathfrak{B}$. Furthermore, the point z is supposed to lie in some fixed domain $\mathfrak{G}$ of the complex plane. We wish to find out under what additional assumptions the sequence of functions

$$\wp_k(z) = \wp(z; \ \omega_{1k}, \ \omega_{2k}) \qquad (k = 1, 2, \ldots)$$

converges uniformly on $\mathfrak{G}$ to a limit function $y(z)$ which is regular apart from poles.

First, we consider the case when the first basic period converges to 0 for some subsequence of the indices k. If z_0 is a point in $\mathfrak{G}$, then, for $k \to \infty$, the points $z_0 + l\omega_{1k}$ $(l = 0, \pm 1, \pm 2, \ldots)$ would accumulate at every point of a certain line through z_0 and, in view of the periodicity of the $\wp$-function and the assumed uniform convergence of the sequence of functions $\wp_k(z)$, the limit function $\varphi(z)$ would be constant on the above line through z_0. As a meromorphic function, $\varphi(z)$ would therefore be identically constant. We shall show that this case cannot occur. We have

(1) $$\frac{\omega_1^4}{60} g_2 = \sum_{m,n}{}' (m + n\tau)^{-4}, \qquad \frac{\omega_1^6}{140} g_3 = \sum_{m,n}{}' (m + n\tau)^{-6},$$

where m, n vary over all pairs of integers with the exception of the pair $0, 0$. The inequality

$$2\,|m + n\tau|^2 - |m + ni|^2 = m^2 + 4\xi mn + (2\xi^2 + 2\eta^2 - 1)n^2$$
$$= (m + 2\xi n)^2 + (2\eta^2 - 2\xi^2 - 1)n^2 \geq 0$$

implies that the series in (1) converge uniformly on $\mathfrak{B}$. If the period ratios $\tau = \tau_k$ corresponding to our subsequence $\omega_1 = \omega_{1k}$ of the first periods had a limit point on $\mathfrak{B}$, then a subsequence of the second periods $\omega_2 = \omega_{2k}$ would converge to 0 together with the ω_{1k} and the corresponding period lattices would be arbitrarily fine. But then for $k \to \infty$ the poles of the functions $\wp_k$ would accumulate about every point of $\mathfrak{G}$, which is impossible. It follows that the τ_k tend to ∞. In view of the uniform convergence of the series, the right-hand sides of (1) tend to the positive limits

(2)
$$2 \sum_{m=1}^{\infty} m^{-4} = 2\zeta(4), \qquad 2 \sum_{m=1}^{\infty} m^{-6} = 2\zeta(6),$$

where, as usual, $\zeta(s)$ denotes the Riemann Zeta-function. Now we make use of the differential equation

(3)
$$\left(\frac{d\wp(z)}{dz}\right)^2 = 4[\wp(z)]^3 - g_2\wp(z) - g_3,$$

and note that for $k \to \infty$ the derivative $[d\wp_k(z)/dz]$ must tend to 0, since the $\wp_k(z)$ converge uniformly to a constant on $\mathfrak{G}$. However, if we multiply (3) by ω_1^6, then, in view of (1) and (2), all terms except the last become 0 as $k \to \infty$. This yields the required contradiction.

We can now assume that for a fixed c

$$|\omega_{1k}| > c > 0 \qquad (k = 1, 2, \ldots).$$

In addition let $|z| < r$ in the domain $\mathfrak{G}$. Then for

$$m^2 + n^2 > 8c^{-2}r^2, \qquad \omega = m\omega_{1k} + n\omega_{2k},$$

we have

$$2\,|z| < \frac{c}{\sqrt{2}}\,|m + ni| \leq c\,|m + n\tau_k| < |\omega|,$$

$$|(z - \omega)^{-2} - \omega^{-2}| = \frac{|(z - 2\omega)z|}{|z - \omega|^2\,|\omega|^2} < \frac{\frac{5}{2}r\,|\omega|}{\frac{1}{4}\,|\omega|^4}$$

$$= 1 \quad \text{or} \quad |\omega|^{-3} < 30rc^{-3}\,|m + ni|^{-3},$$

so that the series of partial fractions of the function $\wp(z;\ \omega_{1k},\ \omega_{2k})$ converges uniformly in z and k. The following three possibilities must now be investigated. Either the sequence ω_{2k} has a finite limit point, or the sequence ω_{2k}

tends to ∞ and the sequence ω_{1k} has a finite limit point, or both sequences tend to ∞. In the first case we may assume, after possibly shifting to an appropriate infinite subsequence, that ω_{1k} and ω_{2k} converge to nonzero limits ω_1 and ω_2, in which case the limit function $\varphi(z)$ is just the function $\wp(z;\ \omega_1,\ \omega_2)$ and so not a degenerate form of the $\wp$-function. In the second case we may assume that ω_{1k} tends to a nonzero limit ω_1 and, in addition, τ_k tends to ∞. In view of the uniform convergence of the series of partial fractions, we have

$$\varphi(z) = \varphi(z;\ \omega_1) = z^{-2} + \sum_m{}' [(z - m\omega_1)^{-2} - (m\omega_1)^{-2}]$$

$$= -2\omega_1^{-2}\zeta(2) + \sum_{m=-\infty}^{\infty} (z - m\omega_1)^{-2}.$$

On the other hand, we have the partial fraction expansion

$$\pi \operatorname{ctg} \pi z = z^{-1} + \sum_{m=-\infty}^{\infty}{}' \{(z - m)^{-1} + m^{-1}\}.$$

Since it converges uniformly, this series admits termwise differentiation. As a result, we obtain the relation

$$\left(\frac{\pi}{\sin \pi z}\right)^2 = z^{-2} + \sum_{m=-\infty}^{\infty}{}' (z - m)^{-2} = \sum_{m=-\infty}^{\infty} (z - m)^{-2}.$$

Subtracting z^{-2} from both sides of this relation and letting z tend to 0, we obtain the formula first proved by Euler

$$2\zeta(2) = \lim_{z \to 0} \left[z^{-2}\left(1 - \frac{\pi^2}{6}z^2 + \cdots\right)^{-2} - z^{-2}\right] = \frac{\pi^2}{3}.$$

It follows that in this case the limit function has the form

$$\varphi(z;\ \omega_1) = \left(\frac{\pi}{\omega_1}\right)^2 \left(\sin\frac{\pi z}{\omega_1}\right)^{-2} - \frac{1}{3}\left(\frac{\pi}{\omega_1}\right)^2.$$

In the third case ω_{1k} and ω_{2k} tend to ∞, so that the limit function is

$$\varphi(z) = z^{-2}.$$

As illustrations of the second and third case we can take the sequences of functions $\wp_k = \wp(z;\ \omega_1,\ ki\omega_1)$ and $\wp_k = \wp(z;\ k,\ ki)$; in either case, we can choose for $\mathfrak{G}$ any bounded domain in the z-plane. All in all, the degenerate forms of the $\wp$-function are the simply periodic functions $\varphi(z;\ \omega_1)$ with basic period ω_1 and the nonperiodic function $\varphi(z) = z^{-2}$, which may be thought of as arising from $\varphi(z;\ \omega_1)$ via $\omega_1 \to \infty$.

Apart from a difference in notation the function $\varphi(z;\ \omega_1)$ appears at the end of Section 13 and arises from the function $q(w)$ in Section 13 by putting $w = z$ and $a = \pi\omega_1^{-1}$. In the differential equation for $q(w)$ we had

$$g_2 = \tfrac{4}{3}a^4 = \frac{4}{3}\left(\frac{\pi}{\omega_1}\right)^4, \qquad g_3 = \tfrac{8}{27}a^6 = \frac{8}{27}\left(\frac{\pi}{\omega_1}\right)^6;$$

(1) and (2) yield the formulas

$$\omega_1^4 g_2 = 120\zeta(4), \qquad \omega_1^6 g_3 = 280\zeta(6).$$

This gives

$$\zeta(4) = \frac{\pi^4}{90}, \qquad \zeta(6) = \frac{\pi^6}{945}.$$

The degenerate forms of $\wp'(z)$ are, accordingly,

$$\varphi'(z; \omega_1) = -2\left(\frac{\pi}{\omega_1}\right)^3\left(\sin\frac{\pi z}{\omega_1}\right)^{-3}\cos\frac{\pi z}{\omega_1}, \qquad \varphi'(z) = -2z^{-3},$$

and we have

$$\frac{2\pi i}{\omega_1}\frac{\varphi(z; \omega_1) + \dfrac{1}{3}\left(\dfrac{\pi}{\omega_1}\right)^2}{\varphi'(z; \omega_1)} = -i\,\mathrm{tg}\frac{\pi z}{\omega_1} = \frac{2}{1 + e^{(2\pi i/\omega_1)z}} - 1,$$

$$-2\frac{\varphi(z)}{\varphi'(z)} = z.$$

It follows that the exponential function

$$\psi(z; \omega_1) = e^{(2\pi i/\omega_1)z}$$

can be expressed rationally in terms of the functions $\varphi(z; \omega_1)$ and $\varphi'(z; \omega_1)$; and, clearly, $\varphi(z; \omega_1)$ and $\varphi'(z; \omega_1)$ are also rational functions of $\psi(z; \omega_1)$. A corresponding result holds for $\varphi(z)$ and the function

$$\psi(z) = z.$$

Finally we note that, for given periods ω_1 and ω_2, all elliptic functions $E(z)$ are given by the rational functions of $\wp(z; \omega_1, \omega_2)$ and $\wp'(z; \omega_1, \omega_2)$ with constant coefficients. If, with these coefficients kept fixed, we put $E(z) = E(z; \omega_1, \omega_2)$ and allow the periods to degenerate as above, then, corresponding to the two possibilities, we obtain as limit function of $E(z; \omega_1, \omega_2)$ a rational function of $\psi(z; \omega_1)$ or $\psi(z)$. Since $\psi(z; \omega_1)$ and $\psi(z)$ themselves appear as such functions, it follows that degeneration of the field of elliptic functions yields the field of rational functions of $e^{(2\pi i/\omega_1)z}$ or the field of rational functions of z.

This result suggests that it might be possible to develop a theory of meromorphic simply periodic functions in a manner similar to that employed for elliptic functions. One basic difference, however, is that there exist simply periodic functions which are not connected by an algebraic equation. One such example is furnished by the two functions e^z and $e^{(e^z)}$. Therefore the theory of simply periodic functions is not nearly as appealing as the theory of doubly periodic functions. The deeper reason is that the behavior of any function at infinity always requires special considerations. In particular, this remark applies to simply periodic functions, since their period strips are unbounded. By contrast, the doubly periodic functions need only be investigated in a period parallelogram which lies completely in the finite part of the plane.

2

Uniformization

1. Algebraic functions

In order to arrive at the definition of an algebraic function we take as our starting point a polynomial $P(w, z)$ in the variables w and z with constant coefficients. We assume that $P(w, z)$ has positive degree n in w. If we write the terms of $P(w, z)$ in descending powers of w, then we have

$$(1) \qquad P(w, z) = P_0(z)w^n + P_1(z)w^{n-1} + \cdots + P_n(z).$$

Here $P_0(z)$, $P_1(z)$, ..., $P_n(z)$ are polynomials in z with constant coefficients, and $P_0(z)$ is not identically zero. We also assume that $P(w, z)$ is irreducible, that is, it cannot be written as a product of two polynomials with constant coefficients neither of which is a constant. It follows that, in particular, the $n + 1$ polynomials $P_0(z)$, ..., $P_n(z)$ have no common divisor which actually depends on z. If we put, for brevity,

$$\frac{P(w, z)}{P_0(z)} = F(w, z), \qquad \frac{P_k(z)}{P_0(z)} = R_k(z) \qquad (k = 1, \ldots, n),$$

then the expansion

$$(2) \qquad F(w, z) = w^n + R_1(z)w^{n-1} + \cdots + R_n(z)$$

is an n-th degree polynomial in w where w^n has the coefficient 1, and the remaining coefficients are rational functions of z with common denominator $P_0(z)$. It is shown in algebra that $F(w, z)$ viewed as a polynomial in w is irreducible over the field of rational functions of z; in other words, it is not possible to write $F(w, z)$ as a product of two polynomials of lower degree in w whose coefficients are again rational functions of z. Conversely, if we start out with an irreducible polynomial in w (in the latter sense) of the form (2), that is with coefficients $R_1(z), \ldots, R_n(z)$ which are given rational functions of z, then upon multiplication by the least common denominator $P_0(z)$ of these coefficients we obtain once more an irreducible polynomial $P(w, z)$ (in the former sense).

The discriminant of an n-th degree polynomial

$$Q(w) = q_0 w^n + q_1 w^{n-1} + \cdots + q_n$$

90

with indeterminate coefficients $q_0, q_1, \ldots, q_n$ is a polynomial

$$D(Q) = D(q_0, q_1, \ldots, q_n)$$

in $q_0, q_1, \ldots, q_n$ with rational integral coefficients. If $q_0, q_1, \ldots, q_n$ are elements of a specified field K, then, as is well known, the significance of the discriminant is that for $q_0 \neq 0$ it vanishes if and only if the polynomial $Q(w)$ and its derivative

$$Q'(w) = nq_0 w^{n-1} + (n-1)q_1 w^{n-2} + \cdots + q_{n-1}$$

have as common divisor a polynomial with coefficients in K which actually contains w. We apply this to

$$Q(w) = P(w, z), \qquad Q'(w) = P_w(w, z).$$

Then K is the field of rational functions of z with constant complex coefficients and the discriminant

$$D(P) = S(z)$$

is a polynomial in z. If the discriminant were identically 0 in z, then $F(w, z)$ and $F_w(w, z)$ would have a common divisor with coefficients in K which would depend on w; whereas $F(w, z)$ is irreducible and the polynomial $F_w(w, z)$, which is not identically zero, is of lower degree in w than $F(w, z)$.

After these algebraic preliminaries we turn to the question of the extent to which the algebraic equation

(3) $$P(w, z) = 0$$

determines w as a function of the complex variable z. In the following investigation we exclude, at first, the finitely many zeros of $P_0(z)S(z)$ and the point ∞ from the complex sphere $\mathfrak{R}$; the result is the punctured number sphere $\dot{\mathfrak{R}}$. We denote the distinct excluded points (∞ among them) by $z_1, \ldots, z_m$. In view of (1), (3) is an equation in the unknown w with complex coefficients whose degree is exactly n. Since the value of the discriminant $S(z)$ is different from 0, it follows that for a given z the equation (3) has n distinct roots $w_1, \ldots, w_n$. We shall now show that it is possible to group the values $w_1, \ldots, w_n$ obtained by varying z, at first locally, into analytic function elements. To do this we shall make use of the well-known implicit function theorems of the differential calculus.

If we put $w = u + iv, z = x + iy$, with independent real variables u, v, x, y, then we have

(4) $$P(w, z) = G(u, v; x, y) + iH(u, v; x, y),$$

where G and H are polynomials with real coefficients. We must now compute the unknowns u and v from the two equations $G = 0$ and $H = 0$ as functions of x and y. For a fixed point $z_0 = x_0 + iy_0$ in $\dot{\mathfrak{R}}$ the equation $P(w, z_0) = 0$ has exactly n distinct roots $w = w_{10}, w_{20}, \ldots, w_{n0}$, and for these values the derivative $P_w(w, z) \neq 0$. Let $w_0 = u_0 + iv_0$ be one of these roots. If we could show that the functional determinant

$$\frac{d(G, H)}{d(u, v)} = G_u H_v - G_v H_u,$$

viewed as a function of u, v, x, y, is different from 0 at $u = u_0, v = v_0, x = x_0$, $y = y_0$, then we could assert the existence, in a sufficiently small real neighborhood $\mathfrak{U}$ of the point x_0, y_0, of exactly one pair of continuously differentiable real functions $u(x, y)$, $v(x, y)$, which take on the values u_0, v_0 at the point $x = x_0, y = y_0$, and satisfy the two equations

(5) $G[u(x, y), v(x, y); x, y] = 0,$ $H[u(x, y), v(x, y); x, y] = 0$

identically in x and y on $\mathfrak{U}$. In order to determine the functional determinant we note that for a fixed $z = z_0$ the polynomial $P(w, z_0)$ is an analytic function of w and as such satisfies the Cauchy-Riemann equations. We therefore have the relations

$$G_u = H_v, \qquad G_v = -H_u, \qquad P_w = P_u = G_u + iH_u,$$
$$G_u H_v - G_v H_u = G_u^2 + H_u^2 = |G_u + iH_u|^2 = |P_u|^2.$$

Now $P_w(w_0, z_0) \neq 0$ proves that the functional determinant does not vanish, and this justifies our use of the above existence theorem.

Next we show that the two functions $u = u(x, y)$, $v = v(x, y)$, which are continuously differentiable in x and y on $\mathfrak{U}$, satisfy themselves the Cauchy-Riemann differential equations provided that $\mathfrak{U}$ is sufficiently small. The relations (4) and (5) show that the equation (3) holds for the complex function $w = w(x, y) = u + iv$ identically in x and y on $\mathfrak{U}$. Since w is continuously differentiable and $P(w, z)$ is a polynomial in w and z, we obtain by differentiation

$$P_w w_x + P_z z_x = \frac{\partial P}{\partial x} = 0, \qquad P_w w_y + P_z z_y = \frac{\partial P}{\partial y} = 0,$$

$$P_w w_x + P_z = 0, \qquad\qquad P_w w_y + iP_z = 0,$$

and, after elimination of P_z,

$$(w_x + iw_y)P_w = 0.$$

Since $P_w(w_0, z_0) \neq 0$, it follows that $P_w(w, z)$, with $w = w(x, y)$, is different from zero in every sufficiently small neighborhood $\mathfrak{U}$ of the point x_0, y_0. But then

$$w_x = -iw_y.$$

Separating the real and imaginary parts in the last relation we obtain

$$u_x = v_y, \qquad v_x = -u_y.$$

This shows that w, as a function of the complex variable $z = x + iy$, is regular analytic in the vicinity of z_0.

We have found n different analytic function elements $w_k(z)$ $(k = 1, \ldots, n)$ which take on the values w_{k0} at $z = z_0$ and satisfy in a neighborhood of z_0 the equation

$$P[w_k(z), z] = 0$$

identically in z. The corresponding power series

$$w_k(z) = w_{k0} + w_{k1}(z - z_0) + w_{k2}(z - z_0)^2 + \cdots$$

are all convergent in a sufficiently small disk $|z - z_0| < r$ on $\mathfrak{R}$. Since the constant terms w_{k0} of these series are distinct, it follows that for r small enough and for every z in the disk the n values $w_k(z)$ are likewise distinct. On the other hand, for every fixed z in $\dot{\mathfrak{R}}$ the equation $P(w, z) = 0$ has exactly n solutions $w_1, \ldots, w_n$. But then, in a small enough neighborhood of z_0 on $\dot{\mathfrak{R}}$, the series $w_k(z)$ $(k = 1, \ldots, n)$ give all the solutions of the equation $P(w, z) = 0$. Now we propose to investigate the analytic continuations of these power series initially excluding poles, that is, using the term analytic continuation in its original sense.

Theorem 1: The power series $w_k(z)$ $(k = 1, \ldots, n)$ can be continued along every path on the punctured sphere $\dot{\mathfrak{R}}$ issuing from z_0, and every function element $w = w(z)$ obtained in the process of analytic continuation satisfies the equation $P(w, z) = 0$.

Proof: If it were not possible to continue the series $w_k(z)$ on $\dot{\mathfrak{R}}$ along every path issuing from z_0, then there would exist a curve C on $\dot{\mathfrak{R}}$ with end point a and beginning point z_0 such that $w_k(z)$ could be continued along C to every point b preceding a, but not to a itself. On the other hand, since z_0 can be chosen arbitrarily on $\dot{\mathfrak{R}}$ there are, in a small enough disk $|z - a| < s$, altogether n distinct power series $W(z) = W_l(z)$ $(l = 1, \ldots, n)$ in the variable $(z - a)$ which satisfy there the equation $P[W(z), z] = 0$, and yield all of its solutions. We now choose b so that the subarc of C between b and a lies entirely in our disk, and consider the function element $w(z)$ which is the result of analytic continuation of the power series $w_k(z)$ from z_0 to b along C. Since $P(w_k(z), z) = 0$ we also have, by analytic continuation, $P[w(z), z] = 0$. But then $w(z)$ must coincide with one of the functions $W_l(z)$ at each point in the disk $|z - b| < s - |b - a|$. The function element $W_l(z)$, however, can be continued from b to a along C, and this contradicts the initial assumption about $w_k(z)$. It follows that the power series $w_k(z)$ can be continued analytically on $\dot{\mathfrak{R}}$ along every path, and every continuation satisfies the equation $P(w, z) = 0$.

The fact that the power series $w_k(z)$, which we found in the vicinity of z_0, can be continued analytically on $\dot{\mathfrak{R}}$ from z_0 to every point does not at all imply independence of the analytic continuation from the path. In order to investigate the multiple-valuedness of $w_k(z)$ in the large and to construct the Riemann surface of $w_k(z)$, we shall first make the punctured sphere $\dot{\mathfrak{R}}$ into a simply connected surface by the addition of appropriate cuts. To do this it is advantageous to add to the excluded points $z_1, \ldots, z_m$ an additional point z^* different from the points $z_1, \ldots, z_m$ and z_0. We denote the resulting punctured sphere by $\mathfrak{R}^*$. We now join the point z^* to the point z_j $(j = 1, \ldots, m)$ by means of a piecewise smooth curve L_j, which is on the sphere $\mathfrak{R}$ and is free of double points, and we stipulate that the m curves L_j do not pass through z_0 and are disjoint except for the point z^*. By dissecting $\mathfrak{R}^*$ along the curves L_j $(j = 1, \ldots, m)$ we obtain a surface $\mathfrak{S}$ with a boundary. We observe that each cut along an L_j contributes two edges to $\mathfrak{S}$ and that the points $z^*, z_1, \ldots, z_m$ do not belong to $\mathfrak{S}$. If we go around the point z^* in a negative sense on a sufficiently small circle Q, then we encounter each of the piecewise smooth curves exactly once, and the curves can be numbered so that they occur in the order $L_1, L_2, \ldots, L_m$ (Figure 33).

Theorem 2: Each of the function elements $w_k(z)$ can be uniquely continued on the surface $\mathfrak{S}$ from z_0 to any point on $\mathfrak{S}$.

Proof: According to the monodromy theorem it suffices to show that $\mathfrak{S}$ is simply connected. Each cut L_j contributes to $\mathfrak{S}$ two edges from z^* to z_j which we shall denote by A_j and B_j. We assume that $\mathfrak{S}$ lies relative to B_j just as the upper half plane lies relative to the real axis, that is, $\mathfrak{S}$ is to the left of B_j. The curve $B_1 A_1^{-1} B_2 A_2^{-1} \cdots B_m A_m^{-1}$ bounds the surface $\mathfrak{S}$ in a positive

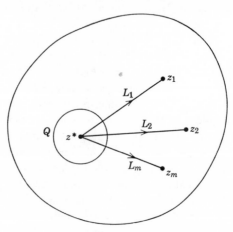

Figure 33

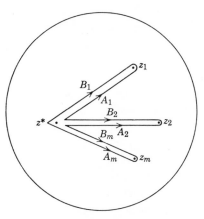

Figure 34

sense (Figure 34). By actually separating the edges of the cuts we can deform $\mathfrak{S}$ continuously into a hemisphere and then into a plane circular disk. It follows that $\mathfrak{S}$ is simply connected as asserted.

We now omit the edges B_j $(j = 1, \ldots, m)$ and call the resulting surface $\mathfrak{S}'$. In view of Theorem 2, the function $w_k(z)$ is uniquely defined on $\mathfrak{S}'$ in the large by analytic continuation; in particular $w_k(z)$ is uniquely defined on the edge A_j. If we restore the edge B_j to $\mathfrak{S}'$, then analytic continuation of $w_k(z)$ on $\mathfrak{S}$ to the points of B_j yields at these points a definite function $w = w(z)$, which also satisfies the equation $P(w, z) = 0$. Now, A_j and B_j coincide on $\mathfrak{R}$, and the totality of solutions of $P(w, z) = 0$ on A_j is given by the n functions $w_1(z), \ldots, w_n(z)$. It follows that $w(z) = w_l(z)$, where the index l depends on k and j and $w_l(z)$ is defined on A_j in a manner analogous to $w_k(z)$. Conversely, if, beginning at B_j, we continue the function $w_l(z)$ on $\mathfrak{S}$ to A_j, then we return to $w_k(z)$. This shows that for a fixed j the correspondence $k \leftrightarrow l_k$ is a permutation of the integers $1, 2, \ldots, n$ which depends on j. We denote this permutation by π_j $(j = 1, \ldots, m)$—this is not to imply that the permutations π_j need all be distinct.

We now construct a Riemann region $\mathfrak{R}$ in the following manner: We take n copies $\mathfrak{S}_1, \mathfrak{S}_2, \ldots, \mathfrak{S}_n$ of the surface $\mathfrak{S}$, which was the result of making the m cuts $L_1, L_2, \ldots, L_m$ in the punctured sphere $\mathfrak{R}^*$, and omit in each of them the edges $B_1, \ldots, B_m$. With the resulting sheet $\mathfrak{S}'_k$ $(k = 1, \ldots, n)$ we associate the single-valued function $w_k(z)$ which is regular there. For every $j, j = 1, \ldots, m$, we join the edge B_j of $\mathfrak{S}_k$ $(k = 1, \ldots, n)$ to the edge A_j of $\mathfrak{S}_l$ $[l = l_k = \pi_j(k)]$. In this way all edges are joined in pairs. We shall see later that the resulting surface $\mathfrak{R}$ is connected. At this point, however, we cannot dismiss the possibility that our surface $\mathfrak{R}$ consists of a number of disjoint parts each of which is connected and made up of certain of the sheets

$\mathfrak{S}_1', \ldots, \mathfrak{S}_n'$. What is clear is that there are exactly n points of $\mathfrak{R}$ over every point z of $\mathfrak{R}^*$. Now we consider an arbitrary closed curve C on $\mathfrak{R}$ issuing from the point z_0 of the sheet $\mathfrak{S}_k'$. The analytic continuation of $w_k(z)$ along C leads back to the same function element at z_0. This is so because a certain terminal arc of C necessarily belongs to $\mathfrak{S}_k'$, where the function is uniquely determined by $w_k(z)$. Observe that when we cross an edge of $\mathfrak{S}_k$ into the adjoining sheet $\mathfrak{S}_l$, $w_k(z)$ goes over by analytic continuation into the function $w_l(z)$ associated with $\mathfrak{S}_l$. This implies that $w_k(z)$ can be uniquely continued to all points of $\mathfrak{R}$ which can be joined by a curve to the point z_0 of $\mathfrak{S}_k'$.

We shall now investigate $\mathfrak{R}$ at the hitherto excluded points $z^*, z_1, \ldots, z_m$, and then complete $\mathfrak{R}$ to a surface $\mathfrak{R}$, which will turn out to be the required Riemann surface of the algebraic function defined by (3). The point z^* presents no difficulties. Specifically, consider the small circle Q on $\mathfrak{R}$ which goes around the point z^* in the negative sense and intersects each of the curves $L_1, \ldots, L_m$ exactly once. These curves break up the circular disk into m sectors and these, in turn, appear on each of the sheets $\mathfrak{S}_1', \ldots, \mathfrak{S}_n'$. If we start with one of the sectors on $\mathfrak{S}_k'$, then, in view of Theorem 1 and the monodromy theorem, the function $w_k(z)$ which is defined there can be uniquely continued to all points of the circular disk including z^*. It follows that after going around Q we must return to the initial function $w_k(z)$. On the other hand, when we cross L_j on Q, then the indices of the $w_k(z)$ ($k = 1, \ldots, n$) are subject to the permutation π_j; hence the product

$$\pi_1 \pi_2 \cdots \pi_m = \varepsilon$$

is the identity permutation. When the sheets are joined, there are n groups of m sectors and each group forms a circular disk punctured at z^*. Since the $w_k(z)$ can be uniquely continued analytically to the full circular disk, we add to $\mathfrak{R}$ the n points over z^*.

Investigation of the points z_j ($j = 1, \ldots, m$) is less simple. Let Q_j be a positively oriented circle about the point z_j which intersects L_j once but does not intersect any of the remaining curves (Figure 35). Suppose that the permutation π_j carries some number k_1 belonging to the sequence $1, 2, \ldots, n$ into k_2, k_2 into $k_3, \ldots, k_{r-1}$ into k_r, and k_r into k_1, where $k_1, k_2, \ldots, k_r$ are distinct.

If, starting with a point of the sheet $\mathfrak{S}_k$, we let z go around the circle Q_j once, then the functions $w_k(z)$ corresponding to $k = k_1, k_2, \ldots, k_r$ are permuted cyclicly, and the same is true of the associated sheets. In the neighborhood of z_j these sheets form an r-sheeted ramified surface or a schlicht neighborhood according as $r > 1$ or $r = 1$. If $r < n$ then we can split off from $\mathfrak{S}_j$ another cycle $l_1, l_2, \ldots, l_s$ and, corresponding to this cycle, we obtain another s-sheeted ramified surface about z_j. If π_j is the product of q disjoint cycles of order $r_1, \ldots, r_q$, then we obtain over z_j q ramified

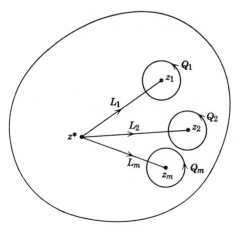

Figure 35

surfaces consisting of $r_1, \ldots, r_q$ sheets, respectively, and $r_1 + r_2 + \cdots + r_q = n$. It should be pointed out that the cycles may depend on j and that our argument is valid only in a sufficiently small neighborhood of z_j. It remains to see whether the points z_j $(j = 1, \ldots, m)$ should be adjoined to the various ramified surfaces as branch points, and this requires the study of the functions $w_k(z)$ for $z \rightarrow z_j$. We assume at first that $z_j \neq \infty$. Let $\Re_j$ denote the circular region in the z-plane bounded by Q_j, and let μ be an upper bound for the absolute values of the n polynomials

$$[P_0(z)]^{k-1}P_k(z) = T_k(z) \qquad (k = 1, \ldots, n)$$

on $\Re_j$; also, let

$$\rho = n\mu + 1.$$

If z is a point of $\Re_j$ and w is a complex number such that

$$|P_0(z)w| \geq \rho,$$

then for

$$v = P_0(z)w$$

we have the inequalities $|v| \geq \rho > 1$ and

$$|[P_0(z)]^{n-1}P(w, z)| = |v^n + T_1(z)v^{n-1} + \cdots + T_n(z)|$$
$$\geq \rho^n(1 - n\mu\rho^{-1}) = \rho^{n-1} > 0.$$

We see that for every solution w of the equation $P(w, z) = 0$ we must have $|v| < \rho$. This implies the boundedness of the function

$$v_k(z) = P_0(z)w_k(z)$$

for $z \to z_j$. If the index k belongs to an r- cycle of the permutation π_j, then we put

$$z = z_j + t^r.$$

As a function of t, $w_k(z)$ admits unique analytic continuation in the neighborhood of $t = 0$ for $t \neq 0$, and the same is true of $v_k(z)$ whose boundedness implies its regularity at the point $t = 0$ itself. On the other hand, the rational function $(P_0(z))^{-1}$ has at $t = 0$ at most a pole; it has a pole if and only if $P_0(z_j) = 0$. It follows that in the neighborhood of $t = 0$ we have a convergent power series expansion

$$(6) \qquad\qquad w_k(z) = \sum_{l=-h}^{\infty} c_l t^l, \qquad t = (z - z_j)^{1/r},$$

and the r different values of the root lead to the r branches of $w_k(z)$ for $k = k_1, \ldots, k_r$. Since in $\Re_j$ the r values of $w_k(z)$ for $z \neq z_j$ are distinct, it follows that, for a fixed k, $k = k_1, \ldots, k_r$, the value of t is uniquely determined in the vicinity of $t = 0$ for $t \neq 0$ by prescribing z and $w_k(z)$. It should be noted that negative powers of t can actually appear in (6) only if $P(z_j) = 0$, and that otherwise $w_k(z)$, as a function of the local parameter t, is regular also at $t = 0$ and the constant term $c_0 = w_k(z_j)$.

To obtain the corresponding result for $z_j = \infty$ we make the substitution $z^{-1} = u$ and put

$$u^g P(w, z) = P^*(w, u), \qquad u^g P_l(z) = P_l^*(u), \qquad (l = 0, \ldots, n),$$

where g is the largest of the degrees of the $n + 1$ polynomials $P_l(z)$, that is, the degree of $P(w, z)$ in z. We can then apply the previous considerations literally to the starred polynomials, where we need only replace z and z_j by u and 0. We then obtain a power series expansion (6) with $t = z^{-(1/r)}$.

These power series expansions are known as *Puiseux series*. They were first constructed by Puiseux in the middle of the 19th century. Without going into the matter we wish to point out that deeper algebraic investigation yields a method for the computation of all the coefficients c_l. Since for $z \to z_j$ the functions $w_k(z)$ approach a finite or infinite limit, z_j must be added as a branch point to each of the q ramified surfaces above it. In this way there arises out of $\Re$ the complete Riemann surface $\Re$ of the algebraic function $w = w(z)$ defined by the equation $P(w, z) = 0$.

It might be conjectured that for $z \neq \infty$ and $P_0(z_j) \neq 0$ the numbers $r_1, \ldots, r_q$ coincide with the multiplicities of the different roots w of the algebraic equation $P(w, z_j) = 0$, so that, in particular, q is the number of different roots. Simple examples show, however, that this is not in general the case. Nevertheless, the identity

$$P(w, z) = P_0(z) \prod_{k=1}^{n} [w - w_k(z)]$$

implies, for $z = z_j$, that the n roots of this equation are $w = w_k(z)$ ($k = 1, \ldots, n$). It follows that if one such root $w_k(z_j) = c_0$ has multiplicity p, then there are at $z = z_j$ exactly p Puiseux series with constant term c_0. But it is possible for these series to belong to different cycles $k_1, \ldots, k_r$, and only the sum of the corresponding orders r to yield the value p. Incidentally, at least one of the roots is a multiple root, for the discriminant $S(z)$ vanishes at $z = z_j$. Nevertheless, it can happen that $q = n$ and $r_1 = r_2 = \cdots = r_n = 1$, so that all the sheets of $\Re$ lie schlicht over z.

Theorem 3: The surface $\Re$ is connected.

Proof: We assume that $\Re$ is not connected. Then it is possible to number the sheets of $\Re$ so that for a certain $h < n$ the sheets $\mathfrak{S}_1, \ldots, \mathfrak{S}_h$ are connected with each other but with none of the remaining sheets. Such a decomposition of $\Re$ into at least two disjoint parts is reflected in the permutation group Π generated by $\pi_1, \ldots, \pi_m$ as follows. If, in the previous notation, the permutation π_j carries the indices $k = 1, \ldots, n$ into $l_1, \ldots, l_n$, then $l_1, \ldots, l_h$ is a permutation of $1, \ldots, h$ and $l_{h+1}, \ldots, l_n$ is a permutation of $h + 1, \ldots, n$. More generally, every element of the group Π is the product of a permutation of the numbers from 1 to h and a permutation of the numbers $h + 1$ to n. A group of permutations of the numbers from 1 to n is called *intransitive* when for at least one natural number $h < n$ certain h of the numbers are invariably permuted among themselves. Otherwise Π is called *transitive*. The disconnectedness of $\Re$ implies the intransitivity of Π and, as is easily shown, the converse assertion is also true.

We introduce an indeterminate w and form the product

$$G(w, z) = (w - w_1)(w - w_2) \cdots (w - w_h),$$

where $w_k = w_k(z)$ ($k = 1, \ldots, n$) stands for the uniquely determined function of z on $\mathfrak{S}$. Then

$$G(w, z) = w^h - E_1 w^{h-1} + E_2 w^{h-2} - \cdots + (-1)^h E_h,$$

where E_l, $l = 1, \ldots, h$, denotes the l-th elementary symmetric polynomial in $w_1, \ldots, w_h$. We shall show that the $E_l = E_l(z)$ are rational functions of the variable z. The function $E_l(z)$ is, in any case, single valued and regular on $\mathfrak{S}$. When we cross a cut L_j the functions $w_1(z), \ldots, w_h(z)$ are permuted among themselves, so that $E_l(z)$ has the same values at corresponding points of the edges A_j and B_j. It follows that $E_l(z)$ is single valued and regular on $\Re^*$. When continued on $\mathfrak{S}$, $w_k(z)$ is regular at the boundary point z^* and the same must therefore be true for $E_l(z)$. Finally let us introduce the appropriate Puiseux series for $w_k(z)$ in the neighborhood of z_j. Here we note that, relative to the permutation π_j, the numbers from 1 to h may form a number of cycles. If $r_1, \ldots, r_g$ are the corresponding numbers of sheets in the vicinity of z_j, then we form their least common multiple r and obtain by means of the substitution

$$z = z_j + s^r (z_j \neq \infty), \qquad z = s^{-r}(z_j = \infty)$$

a Laurent series in the variable s for $E_l(z)$ in the neighborhood of z_j on $\dot{\Re}$ which contains only finitely many negative powers. It follows that at $z = z_j$ either $E_l(z)$ has a pole or is regular. But then, apart from a finite number of poles, our function is regular on the full sphere $\Re$ and so is a rational function of z as asserted.

Let $Q_0(z)$ denote the least common denominator of the h rational functions

$$(-1)^l E_l(z) = \frac{Q_l(z)}{Q_0(z)} \qquad (l = 1, \ldots, h),$$

which is determined up to a nonzero constant factor. If we put

$$Q_0(z)G(w, z) = Q(w, z),$$

then

$$Q(w, z) = Q_0(z)w^h + Q_1(z)w^{h-1} + \cdots + Q_h(z)$$

is a polynomial in w and z which is precisely of degree $h < n$ in w. It is well known that two polynomials

$$L = \lambda_0 w^n + \lambda_1 w^{n-1} + \cdots + \lambda_n, \qquad M = \mu_0 w^h + \mu_1 w^{h-1} + \cdots + \mu_n$$

with indeterminate coefficients $\lambda_0, \ldots, \lambda_n$ and $\mu_0, \ldots, \mu_h$ and their resultant $R = R(L, M)$ satisfy an identity

(7) $R = UL + VM,$

where U and V are polynomials of degree $h - 1$ and $n - 1$ in w with coefficients which are again polynomials in $\lambda_0, \ldots, \lambda_n$ and $\mu_0, \ldots, \mu_h$ with integer coefficients and R is free of w. If K is any field containing the λ_k and μ_l, then the vanishing of the resultant when $\lambda_0\mu_0 \neq 0$ signifies that L and M have a common divisor which is a polynomial of degree at least one in w with coefficients in K. In particular, let

$$L = P(w, z), \qquad M = Q(w, z),$$

so that K may be chosen as the field of rational functions of z with complex coefficients. Since $h < n$ and $P(w, z)$ is irreducible, it follows, just as in the previous investigation of the discriminant, that L and M have no such common divisor and $R(L, M)$ is therefore a polynomial in z which does not vanish identically. On the other hand, if we put for w one of the functions $w_1(z), \ldots, w_h(z)$, then both $P(w, z)$ and $Q(w, z)$ vanish identically in z, and (7) yields a contradiction. This proves Theorem 3.

Let $\mathfrak{z}$ denote any point on $\Re$ and z its projection on the number sphere $\Re$. The functions $w_k(z)$ defined on the various sheets form a function $w = w(\mathfrak{z})$

which is defined and meromorphic on all of $\Re$. *This is the algebraic function defined by the equation $P(w, z) = 0$ and $\Re$ is its Riemann surface.* We note that in constructing $\Re$ we made use of the auxiliary point z^* and the branch cuts $L_1, \ldots, L_m$ to introduce the various sheets of $\Re$ and the corresponding branches $w_k(z)$ of w. As we remarked in Section 4, however, the Riemann surface of the function w is independent of our particular construction. If we again omit the branch cuts, then the functions $w_k(z)$ $(k = 1, \ldots, n)$ and the corresponding n sheets are uniquely defined only for a sufficiently small neighborhood of the initial point z_0. Over z_0 there lie n points $\mathfrak{z}_0^{(1)}, \ldots, \mathfrak{z}_0^{(n)}$ corresponding to the n function elements $w_1(z), \ldots, w_n(z)$ given at z_0, and in the course of analytic continuation along every path on $\Re$ joining $\mathfrak{z}_0^{(k)}$ to $\mathfrak{z}_0^{(l)}$ $w_k(z)$ goes over into $w_l(z)$. It is clear that in constructing the Riemann surface we can start with any one of the n power series $w_k(z)$. At this point it should be noted that the numbering of the n sheets is determined by the numbering of the branches $w_1(z), \ldots, w_n(z)$. If the indices $1, \ldots, n$ of the branches are subjected to a permutation λ, then the permutation π_j $(j = 1, \ldots, m)$ must be replaced by $\lambda^{-1}\pi_j\lambda$ and the group Π generated by the π_j must be replaced by the group $\lambda^{-1}\Pi\lambda$. If C is a closed path on the punctured sphere $\dot{\Re}$ issuing from z_0, then in the process of analytic continuation along C the function elements $w_k(z)$ are subject to a permutation and these permutations for various C form the very group Π generated by $\pi_1, \ldots, \pi_m$. The group Π is called the *monodromy group of the algebraic function* $w(\mathfrak{z})$, and it is easy to show that this group is also the Galois group of the algebraic equation $P(w, z) = 0$ with respect to the field of rational functions of z with complex coefficients. The monodromy group will not play any role in the sequel.

We say of a function $f(\mathfrak{z})$ which is uniquely defined at all points of $\Re$ that it is *meromorphic* on $\Re$ if, in terms of the local uniformizing parameter, any point is a pole or a point of regularity of the function. This definition, first given in Section 4, implies that the meromorphic functions on $\Re$ form a field. This field contains the function $w = w(\mathfrak{z})$ and, trivially, the function z itself and, therefore, every rational function of w and z with constant coefficients. We shall show that these are all the functions of the field in question; that is, that this field is obtained by adjoining the algebraic function $w(\mathfrak{z})$ to the field of rational functions of z.

Theorem 4: Every meromorphic function on $\Re$ is of the form

$$f(\mathfrak{z}) = T_1(z)w^{n-1} + T_2(z)w^{n-2} + \cdots + T_n(z), \qquad w = w(\mathfrak{z}),$$

where $T_1(z), \ldots, T_n(z)$ are rational functions of z.

Proof: On the sheet $\mathfrak{S}'_k$ $(k = 1, \ldots, n)$ $f(\mathfrak{z})$ is a function $f_k(z)$ which is regular except for poles and $w_k(z)$ is a simple zero of the polynomial $P(w, z)$, regarded as a polynomial in w. The quotient

$$\frac{P(w, z)}{w - w_k(z)} = Q_k(w, z)$$

is a polynomial of degree $n - 1$ in w which goes over for $w = w_k(z)$ into the function

$$P_w[w_k(z), z] = Q_k[w_k(z), z]$$

of z which is not identically zero. Using the Lagrange interpolation formula we form the expression

$$(8) \qquad g(w, z) = \sum_{k=1}^{n} \frac{f_k(z)P(w, z)}{[w - w_k(z)]P_w[w_k(z), z]}$$

$$= T_1(z)w^{n-1} + T_2(z)w^{n-2} + \cdots + T_n(z);$$

here $T_1(z), T_2(z), \ldots, T_n(z)$ are certain functions of z which, incidentally, are linear expressions in $f_1(z), \ldots, f_n(z)$ with coefficients which are rational expressions in $z, w_1(z), \ldots, w_n(z)$. Let the variable z be restricted to $\mathfrak{S}'$. Now we show that $T_1(z), \ldots, T_n(z)$ are rational functions of z. To this end we substitute a constant u for w. If, beginning at z_0, we continue the function $g(w, z)$ in (8) analytically on the punctured spherical surface $\mathfrak{R}^*$ along any closed path, then this function goes over into itself; this is so because the indices on the branches $w_k(z)$ and on the functions $f_k(z)$ are permuted in the same way, and therefore the sum (8) remains unchanged. Further, using the Puiseux series, we see that the function $g(u, z)$ is either regular at the boundary points $z^*, z_1, \ldots, z_m$ of $\mathfrak{R}^*$ or has poles there. But then $g(u, z)$ is meromorphic on the full number sphere $\mathfrak{R}$, and so is a rational function. If we now choose for w a total of n different constants $u_1, \ldots, u_n$, then (8) yields a system of n linear equations for the functions $T_1(z), \ldots, T_n(z)$ whose determinant is just the Vandermonde determinant of $u_1, \ldots, u_n$, and therefore does not vanish. Since the left sides $g(u_1, z), \ldots, g(u_n, z)$ are known to be rational functions of z, the same is true of $T_1(z), \ldots, T_n(z)$. On the other hand, if we put in (8) $w = w_k(z)$, then we have

$$g[w_k(z), z] = f_k(z),$$

and therefore

$$g[w(\mathfrak{z}), z] = f(\mathfrak{z}),$$

which proves our assertion. Also, the irreducibility of the polynomial $P(w, z)$ shows that the representation of $f(\mathfrak{z})$ given in Theorem 4 is unique.

It is clear that Theorem 7 in Section 14 is a special case of the theorem just proved; we need only choose

$$P(w, z) = w^2 - 4z^3 + g_2z + g_3.$$

Further, the two-sheeted Riemann surfaces constructed in Section 5 of Chapter 1 are special cases of the Riemann surface considered in this section. Although it is true that there the branch cuts were placed differently, this has no effect on the outcome. Now it is natural to ask to what extent it is possible to generalize our earlier results pertaining to the elliptic integral of the first kind and its inverse function by starting with an arbitrary elliptic function. It turns out that one is led in this way to two different classes of problems, both of which are of great significance in the theory of functions.

It is rather obvious how to generalize the notion of an elliptic integral. So far the term elliptic integral stood for an expression of the form

$$s = \int_{z_0}^{z} R(w, z)\, dz, \qquad w^2 = 4z^3 - g_2 z - g_3,$$

where $R(w, z)$ is a rational function of w and z, that is, an arbitrary function analytic on the Riemann surface $\mathfrak{R}$ of w. At the end of Section 14 of the first chapter we showed that in this class of expressions the elliptic integrals of the first kind could be characterized to within an additive and multiplicative constant by the property of being analytically continuable along every path on $\mathfrak{R}$ and staying finite. We showed that the elliptic integral of the first kind maps the covering surface of $\mathfrak{R}$ in a one-to-one conformal manner onto the full s-plane. It also turned out, that as a result of this conformal mapping the field of elliptic functions goes over into the field of functions meromorphic on $\mathfrak{R}$. The generalization of elliptic integrals is given by the expressions

$$s = \int_{z_0}^{z} R(w, z)\, dz, \qquad P(w, z) = 0,$$

where $R(w, z)$ is a rational function of w and z. These expressions are called *abelian integrals*, for Abel was the first to discover one of their most important properties. In analogy to the special case of elliptic integrals we define as *abelian integrals of the first kind* those abelian integrals which can be continued analytically on the Riemann surface $\mathfrak{R}$ of w along every path and stay everywhere finite. In distinction to the elliptic case the corresponding integrand is not, in general, determined up to a constant factor. Specifically, we shall show in the third chapter (Vol. II) that it is possible to choose a minimal number p of fixed integrands $R_1(w, z), \ldots, R_p(w, z)$ such that every integrand in question can be written as a linear combination with suitable complex coefficients in the form

$$R(w, z) = c_1 R_1(w, z) + \cdots + c_p R_p(w, z).$$

The number p is also the genus of the Riemann surface defined geometrically in Section 3; in the elliptic case we have $p = 1$. As we shall see, in the case $p > 1$ an abelian integral of the first kind no longer effects a conformal mapping onto the complex plane, and as a result the inverse function does not yield a function of s which is single valued in the large. Since this single-valuedness played a decisive role in the investigations discussed in Chapter 1, we shall in the rest of this chapter give up the investigation of abelian integrals and, by way of generalization of the Riemann mapping theorem, construct another function s with the desired property. For this it is necessary to give a suitable general definition of the covering surface of $\Re$ and to study the connectivity of $\Re$ in greater detail.

As a result of introducing the uniformizing parameters, the meromorphic functions on $\Re$ go over into *automorphic functions* which are the required analog of the elliptic functions. Their theory is studied in detail in the third chapter (Vol. II). The theory of abelian integrals is developed independently in Chapter 4 (Vol. II). This chapter also contains Abel's theorem, which is a generalization of Euler's addition theorem for elliptic functions. Further, the problem of inversion of an elliptic integral of the first kind is carried over to the the case $p > 1$ in such a way that one arrives nevertheless at single-valued functions. It is necessary, however, to utilize not the inverse function of a single abelian integral of the first kind (which, as we know, is not single valued in the large), but rather the so-called *abelian functions* of p variables introduced by Riemann and Weierstrass. These functions are also a generalization of the elliptic functions, for they are meromorphic, have an addition theorem, and possess $2p$ linearly independent periods. We see that for $p > 1$ there correspond to the elliptic functions two entirely different classes of functions, namely, the automorphic functions of one variable, and the abelian functions of p variables. In Chapter 5 (Vol. III) we develop the general theory of abelian functions and, finally, in Chapter 6 (Vol. III) we discuss the theory of automorphic functions of many variables, with special attention given to modular functions of degree p.

2. Compact Riemann regions

We wish to describe once more the construction of the Riemann surface discussed in the previous section without, however, bringing in the equation $P(w, z) = 0$ of the algebraic function w. To this end we choose $m + 1$ different points $z_1, \ldots, z_m$ and z^* on the complex number sphere, and join z^* to the z_j by means of curves L_j ($j = 1, \ldots, m$) which are free of double points and which meet only at the common beginning z^*. The numbering is chosen so that when we circle z^* in the negative sense, the curves $L_1, \ldots, L_m$ occur in cyclic order. With each curve L_j there is associated a permutation π_j of the numbers 1 to n such that the group of permutations generated by $\pi_1, \ldots, \pi_m$ is transitive and the generators satisfy the condition

$$\pi_1 \cdots \pi_m = \varepsilon,$$

where ε is the identity permutation. We make the m cuts in the number sphere and join n copies of the dissected number sphere along the cuts in the manner indicated by the permutations associated with the cuts. In this way we obtain over the number sphere $\mathfrak{K}$ an n-sheeted surface $\mathfrak{R}$ which, according to the definition in Section 4 of the first chapter, is a Riemann region. Its branch points lie only over the points $z_1, \ldots, z_m$; specifically, for at least one branch point of $\mathfrak{R}$ to lie over z_j it is necessary and sufficient that $\pi_j \neq \varepsilon$. It is again easy to see that the Riemann region $\mathfrak{R}$ does not depend on z^* and the branch cuts L_j but rather on the points $z_1, \ldots, z_m$ and the permutations $\pi_1, \ldots, \pi_m$. We shall say of a Riemann region constructed in this manner that it is of *algebraic type*. This terminology is justified in part by the fact that the class of Riemann regions under consideration includes at least the Riemann surfaces of the algebraic functions. We shall prove much later in Section 8 that, conversely, for every Riemann region of algebraic type there exists an algebraic function whose Riemann surface is the region in question.

Theorem 1: A Riemann region is compact if and only if it is of algebraic type.

Proof: Let $\mathfrak{R}$ be a Riemann region of algebraic type. We must show that every sequence of points $\mathfrak{z}_1, \mathfrak{z}_2, \ldots$ on $\mathfrak{R}$ has a limit point on $\mathfrak{R}$. Let $z^{(k)}$ ($k = 1, 2, \ldots$) be the projection of $\mathfrak{z}_k$ on the number sphere. Since the latter is compact, the sequence of the $z^{(k)}$ has a limit point a. Let $\mathfrak{a}_1, \ldots, \mathfrak{a}_h$ be the points of $\mathfrak{R}$ over a, where h is at most equal to the (finite) number of sheets of $\mathfrak{R}$. At least one of the points $\mathfrak{a}_k$ must be a limit point of the sequence $\mathfrak{z}_1, \mathfrak{z}_2, \ldots$.

Conversely, let $\mathfrak{R}$ be a compact Riemann region. If $\mathfrak{R}$ had infinitely many branch points $\mathfrak{z}_1, \mathfrak{z}_2, \ldots$, then these would have a limit point on $\mathfrak{R}$, which would itself be a branch point or an ordinary point. Both of these possibilities are inadmissible; in fact, a sufficiently small neighborhood of a branch point contains no additional branch points and a sufficiently small neighborhood of an ordinary point contains no branch points at all. It follows that $\mathfrak{R}$ has at most finitely many branch points with distinct projections $z_1, \ldots, z_m$. Let $\dot{\mathfrak{K}}$ be the number sphere punctured at $z_1, \ldots, z_m$. We choose on $\mathfrak{R}$ an ordinary point $\mathfrak{z}_0$ with projection z_0. Now let C be a curve on $\dot{\mathfrak{K}}$ issuing from z_0 with end point z. We wish to show that there is a curve $\mathfrak{C}$ on $\mathfrak{R}$ issuing from $\mathfrak{z}_0$ and having C for its projection. This is certainly true if C is in a sufficiently small neighborhood of z_0; in fact, in this case $\mathfrak{C}$ is uniquely determined. If our assertion were false, then there would be a point b on C such that for every point q_k on C preceding b the subarc from z_0 to q_k would be the projection of a curve $\mathfrak{C}_k$ issuing from $\mathfrak{z}_0$, but the corresponding statement would not hold for b itself. Thus let $q_1, q_2, \ldots$ converge to b and let $\mathfrak{z}_k$ ($k = 1, 2, \ldots$) be the end point of $\mathfrak{C}_k$. The sequence $\mathfrak{z}_1, \mathfrak{z}_2, \ldots$ has on $\mathfrak{R}$ a limit point $\mathfrak{b}$ over b, and $\mathfrak{b}$ is not a branch point. In the limit the curves $\mathfrak{C}_k$ would yield a curve on $\mathfrak{R}$ connecting $\mathfrak{z}_0$ and $\mathfrak{b}$ contrary to the assumption about b. At the same time our argument shows that there is only one curve $\mathfrak{C}$ with the desired property. Since z is an arbitrary number different from $z_1, \ldots z_m$, it follows, in particular, that there is at least one point of $\mathfrak{R}$ over every point of $\dot{\mathfrak{R}}$.

Next we show that the number of points of the Riemann region which lie over a given point of $\dot{\mathfrak{R}}$ is the same for all points of $\dot{\mathfrak{R}}$. An argument similar to that used in proving the finiteness of the number of branch points of $\mathfrak{R}$ shows that there are only finitely many points of $\mathfrak{R}$ over z_0. Let these points be $\tau_1, \ldots, \tau_n$, $n \geq 1$. Over every curve C on $\dot{\mathfrak{R}}$ issuing from z_0 there lie n uniquely determined curves $\mathfrak{C}_1, \ldots, \mathfrak{C}_n$ with initial points $\tau_1, \ldots, \tau_n$. Their terminal points $\mathfrak{z}_1, \ldots, \mathfrak{z}_n$ have the common projection z, the terminal point of C. If these n terminal points were not distinct then, by reversing the direction of C, we would arrive at the contradiction that the n initial points $\tau_1, \ldots, \tau_n$ of the curves $\mathfrak{C}_1, \ldots, \mathfrak{C}_n$ are not all distinct. Also, this argument shows that there are exactly n points of $\mathfrak{R}$ over z.

Let us now puncture $\dot{\mathfrak{R}}$ at an additional point z^* distinct from z_0 and dissect it as before along $L_1, \ldots, L_m$ to obtain again the simply connected surface $\mathfrak{S}$. If C_1 and C_2 are two curves on $\mathfrak{S}$ from z_0 to $z \neq z^*$, then the curve $C_1^{-1}C_2$ is homotopic to zero on $\mathfrak{S}$. This shows that the two curves on $\mathfrak{R}$ over C_1 and C_2 with common initial point τ_k lead to the same point $\mathfrak{z}_k$ over z. If, as before, we go over to the surface $\mathfrak{S}'$ by omitting, in a consistent fashion, one of the two edges of each cut, then we see that there are exactly n sheets $\mathfrak{S}_1', \ldots, \mathfrak{S}_n'$ of $\mathfrak{R}$ over $\mathfrak{S}'$. By circling z_j $(j = 1, \ldots, m)$ in the positive sense we obtain the permutation π_j, which describes the manner in which the cuts over L_j are joined. Since there is no branch point over z^* and $\mathfrak{R}$ is connected, we obtain once more the relation $\pi_1, \cdots \pi_m = \varepsilon$, and the transitivity of the group generated by $\pi_1, \ldots, \pi_m$. This completes our proof.

In Section 5 of the first chapter we mapped the two-sheeted Riemann surface $\mathfrak{R}_4$ topologically onto a torus, and the torus, after canonical dissection, onto a rectangle with identified opposite sides. We now propose to transform an arbitrary compact Riemann region in an analogous manner. Let p be a natural number. We punch $2p$ disjoint circular holes in the spherical surface and join their boundaries in pairs by means of disjoint tubes in space. The resulting closed surface is called a *sphere with p handles*. It is convenient to speak of a spherical surface as a sphere with zero handles. We shall prove

Theorem 2: Every compact Riemann surface is homeomorphic to a sphere with handles.

Proof: Let $\mathfrak{R}$ be a compact Riemann region. In view of Theorem 1 $\mathfrak{R}$ is of algebraic type. We dissect $\mathfrak{R}$ by means of the cuts L_j $(j = 1, \ldots, m)$

into n congruent sheets $\mathfrak{S}_1, \ldots, \mathfrak{S}_n$ (the notation is the same as before). To avoid unnecessary difficulties we use as cuts only arcs of great circles on $\mathfrak{R}$ which issue from a suitably chosen north pole z^* and avoid the south pole. We can assume $m \geq 3$ by possibly including among the z_j the projections of ordinary points. By means of simple topological deformations of $\mathfrak{R}$ we can arrange for the z_j to form a regular m-gon and for the projections of the L_j to be quarter circles issuing from the north pole z^*. The end points z_j and the vertex z^* determine a regular pyramid onto which we project $\mathfrak{R}$ from a point in the interior of the northern hemisphere which lies on the diameter through z^* (Figure 36). Finally by means of suitable individual projections of the sides of the pyramid onto the equatorial plane, we obtain a regular polygon $\mathfrak{P}$ with $2m$ successive sides $C_1, D_1, \ldots, C_m, D_m$ which is the topological image of each of the sheets $\mathfrak{S}$ of $\mathfrak{R}$ with C_j and D_j arising from the edges B_j and A_j^{-1} (Figure 37). Identification of B_j and A_j on $\mathfrak{R}$ is reflected in the congruence correspondence of C_j and D_j^{-1}. For the n sheets of $\mathfrak{R}$ we obtain n regular polygons $\mathfrak{P}_1, \ldots, \mathfrak{P}_n$. Corresponding to the manner in which the edges of the cuts are joined on $\mathfrak{R}$, each of the n sides C_j of these polygons is joined to one of the n sides D_k, with C_j and D_j oppositely oriented; this for $j = 1, \ldots, n$. To obtain a single schlicht polygon in the plane we must admit certain additional topological mappings. If the identification of two corresponding sides is still to be expressible in terms of the relation of congruence, then the only admissible mappings of a polygon are those which reduce to similitudes when considered on each of the sides of the polygon. We shall call such a topological mapping a *transformation*. It is easy to see that two polygons with the same number of sides can be mapped onto each other by means of a transformation. In fact, for two

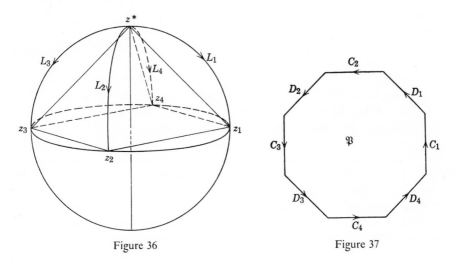

Figure 36   Figure 37

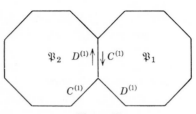

Figure 38

triangles this can be achieved by means of a suitable plane affine transformation and, in the case of arbitrary polygons, the assertion can be proved
inductively by introducing diagonals. If two regular polygons with m_1 and
m_2 equal sides touch externally along a side then, as can be shown by an
induction argument after addition of a diagonal, the resulting simple polygon
with $m_1 + m_2 - 2$ sides can be mapped by means of a transformation onto a
regular polygon of equal side length. Now we proceed to join the polygons
$\mathfrak{P}_1, \ldots, \mathfrak{P}_n$. Since $\mathfrak{R}$ is connected, the numbering can be arranged so that
for $k = 2, \ldots, n$ at least one side $D = D^{(k-1)}$ of $\mathfrak{P}_k$ must be identified with
an oppositely oriented side $C = C^{(k-1)}$ of one of the polygons $\mathfrak{P}_1, \ldots, \mathfrak{P}_{k-1}$.
For $n > 1$ we make $\mathfrak{P}_2$ and $\mathfrak{P}_1$ touch externally along the sides $D^{(1)}$ and $C^{(1)}$
(Figure 38) and transform the resulting figure into a regular polygon $\mathfrak{P}_2^*$ with
$4m - 2$ sides one of which is $C^{(2)}$. For $n > 2$ we make $\mathfrak{P}_3$ and $\mathfrak{P}_2^*$ touch externally along the sides $D^{(2)}$ and $C^{(2)}$ and transform the resulting figure into a regular polygon $\mathfrak{P}_3^*$ of $6m - 4$ sides one of which is $C^{(3)}$. By induction on n we
obtain a regular polygon $\mathfrak{P} = \mathfrak{P}_n^*$ with an even number $2q$ of sides, $2q =
2(nm - n + 1) \geq 6$. The sides form q pairs A, A' where A and A' are
identified and the orientation of A is opposed to that of A'. If we cut $\mathfrak{R}$ along
the corresponding curves, then we obtain a simply connected surface $\mathfrak{R}_0$
homeomorphic to $\mathfrak{P}$. We could therefore already regard the polygon $\mathfrak{P}$ as an
analogue of the period parallelogram in the present general case. It is
possible, however, to effect substantial simplifications by reshaping $\mathfrak{P}$ in a
manner which we are about to discuss.

The positively traversed boundary of $\mathfrak{P}$ is a closed oriented path. By
choosing one of the vertices as an initial point we can describe this path as the
product of the $2q$ sides A, A' taken in a definite order. Since any cyclic
permutation of the factors in this product can be obtained merely by changing
the initial point of the path, we do not distinguish between any two of its $2q$
cyclic variants, each of which is said to define the same word. The $2q$ factors
are viewed as symbols of the word. We shall now subject the given word to
certain changes, and these will be reflected in a change of the system of cuts
on $\mathfrak{R}$.

Suppose that the symbols A, A' of the same pair appear in the word next
to each other. After possibly interchanging the meaning of A and A' we can

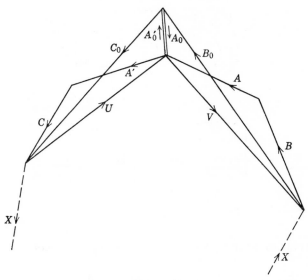

Figure 39

write the word in the form $BAA'CX$, where B and C are symbols and X is a subword consisting of at least one symbol. Using the diagonals U and V we transform the triangles $UA'C$ and VBA into the triangles UA'_0C_0 and VB_0A_0 and then amalgamate A_0 and A'_0 (Figure 39). The resulting polygon B_0C_0X has two sides less than $\mathfrak{P}$ and is carried by a suitable transformation into a regular polygon with which there is associated the abbreviated word BCX. This process of decreasing the length of a word has as its counterpart on $\mathfrak{R}_0$ the joining of two edges, with the surface remaining simply connected. The word BCX may admit further reduction. A reduced word consisting of four symbols in which the symbols A, A' appear next to each other can be put in the form $AA'BB'$ (Figure 40), and then the corresponding square can be deformed to a sphere by joining A' and A^{-1} and B' and B^{-1}. In this case $\mathfrak{R}$ is likewise homeomorphic to a sphere. If the four-symbol word contains no paired off symbols next to each other, then it can be put in the form $ABA'B'$ (Figure 41), and this is precisely the case when $\mathfrak{R}$ is

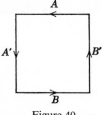

Figure 40

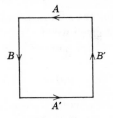

Figure 41

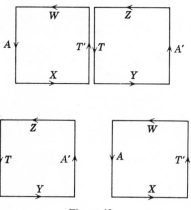

Figure 42

homeomorphic to a torus. Now we need only consider words with at least six symbols.

If the symbols A, A' of a pair appear in a word but are not next to each other, then we may suppose the word to be in the form $WAXYA'Z$, where Y and W are subwords consisting of one or more symbols. In order not to overlook special cases we must allow Z and X to be the empty symbol E, which may be introduced or eliminated at will and has the geometric significance of a point. In the diagram (Figure 42) W, X, Y, Z are represented by means of segments. The polygon is cut along the diagonal T and, after proper transformation of the subpolygon $AXT'W$, this subpolygon and the subpolygon $A'ZTY$ are joined externally along the sides A and A'. After another transformation we obtain a regular polygon $ZTYXT'W$. Formally, the new word is obtained from the old one by replacing A, A' by T, T' and by interchanging X and Y as well as W and Z. On $\mathfrak{R}_0$ the edges corresponding to A and A' are joined, and a new cut is introduced whose edges correspond to the symbols T and T'; at the same time, the surface remains simply connected. We refer to this type of alteration of the given word as *diagonalization* and write

$$WAXYA'Z \simeq ZTYXT'W.$$

We shall now show how to use reduction of length and diagonalization to reduce a given word V to a normal form. Let A_l, A'_l, B_l, B'_l ($l = 1, \ldots, r$) be $2r$ different pairs of symbols and

$$A_l B_l A'_l B'_l = K_l, \qquad K = K_1 K_2 \cdots K_r, \qquad V = KU,$$

where U is a reduced subword, and where we admit the possibility $K = E$ for $r = 0$. We choose in U a pair of symbols C, C' which are as close as possible so that C and C' are separated by at least one symbol D, but not by the symbols D and D'. After possibly renaming and permuting the symbols we can write

$$V = X_1 C X_2 D X_3 C' X_4 D',$$

where K is a subword of X. We put, for brevity,

$$X_1 X_4 X_3 X_2 = PKQ = RQ, \qquad R = PK.$$

After six diagonalizations and a number of cyclic permutations we obtain,

$$\begin{aligned}
V &= X_1 C X_2 (D X_3) C' (X_4 D') \simeq (X_4 D') C_1 (D X_3) X_2 C_1' X_1 \\
&= (X_1 X_4) D' E C_1 D (X_3 X_2 C_1') \simeq (X_3 X_2 C_1') D_1' C_1 E D_1 (X_1 X_4) \\
&= C_1' D_1' C_1 D_1 X_1 X_4 X_3 X_2 = Q C_1' E D_1' C_1 (D_1 R) \\
&\simeq (D_1 R) C_2' D_1' E C_2 Q = C_2' D_1' E (C_2 Q) D_1 R \simeq R D_2' (C_2 Q) E D_2 C_2' \\
&= D_2' C_2 E (Q D_2) C_2' R \simeq R C_3' (Q D_2) E C_3 D_2' = (C_3' Q) D_2 E C_3 D_2' R \\
&\simeq R D_3 C_3 E D_3' (C_3' Q) = K D_3 C_3 D_3' C_3' (Q P).
\end{aligned}$$

Since K is followed by $D_3 C_3 D_3' C_3'$ we see that to justify the transition from r to $r + 1$ in the induction argument we need only note that diagonalization does not affect the length of a word, and that each application of the reduction procedure shortens the length of the subword QP by two.

In this way we obtain the normal form

$$(1) \qquad W_p = K_1 K_2 \cdots K_p, \qquad K_l = A_l B_l A_l' B_l' \qquad (l = 1, \ldots, p),$$

p a positive integer, or the normal form

$$(2) \qquad\qquad\qquad W_0 = A A' B B'.$$

The case (2) was discussed above. We recall that in that case $\mathfrak{R}$ is homeomorphic to a sphere. In case (1), W_p yields a regular polygon $\mathfrak{R}_p$ of $4p$ sides which are identified in pairs in a definite manner. To make this figure into a sphere with p handles we introduce, for $p > 1$, p diagonals $D_1, \ldots, D_p$. The diagonal D_l and the sides A_l, B_l, A_l', B_l' form a pentagon $\mathfrak{F}_l$ ($l = 1, \ldots, p$), which we cut off from $\mathfrak{R}_p$ (Figure 43). By joining A_l to A_l' and B_l to B_l' we make $\mathfrak{F}_l$ into a holed torus $\mathfrak{T}_l$ with the hole bounded by a curve corresponding to the diagonal D_l (Figure 44). With the p pentagons $\mathfrak{F}_l$ cut off, $\mathfrak{R}_p$ reduces to a regular p-gon, which we then deform into a sphere with p holes whose boundaries again come from the diagonals D_l and meet in a single point (Figure 45). By properly attaching the p tori, each with a hole, to the sphere with p holes, we obtain a sphere with p handles which we denote by $\mathfrak{R}_p$ (Figure 46). For $p = 1$, $\mathfrak{R}_p$ stands for the torus, and this surface can be deformed into a sphere with one handle. This completes the proof.

Theorem 2 enables us to visualize the connectivity relations of a compact

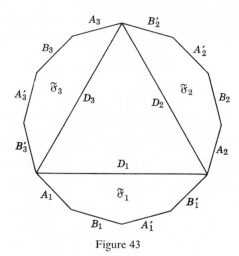

Figure 43

Riemann region. We note that a sphere with handles can be deformed into one of the multiple pretzels mentioned at the end of Section 5 in Chapter 1. In the sequel we shall frequently find it convenient to work not with the sphere with p handles (for $p > 0$) but with the polygon $\mathfrak{N}_p$ whose sides follow one another in the order A_l, B_l, A_l', B_l' for $l = 1, \ldots, p$. The counterpart of the two pairs of sides A_l, A_l' and B_l, B_l' on the torus $\mathfrak{T}_l$ with a hole are two simple closed curves G_l and H_l which intersect in just one point. At a later stage in the combining process these points of intersection on the $\mathfrak{T}_l$ ($l = 1, \ldots, p$) are joined at one point of $\mathfrak{R}_p$. Conversely, $\mathfrak{N}_p$ arises from $\mathfrak{R}_p$ by dissection along the p pairs of crosscuts G_l, H_l ($l = 1, \ldots, p$). On the other hand, the topological mapping associates with every pair of crosscuts G_l, H_l on $\mathfrak{R}_p$ a pair of crosscuts U_l, V_l on the Riemann region $\mathfrak{R}$. These $2p$ cuts issue from

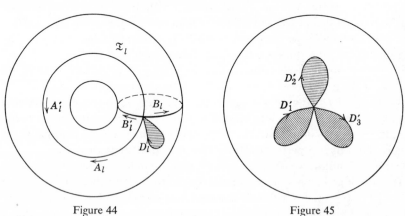

Figure 44   Figure 45

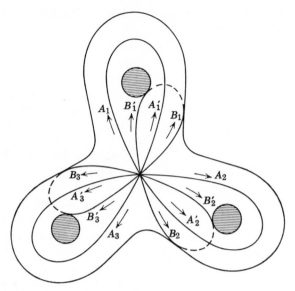

Figure 46

a single point $\mathfrak{y}$ of the Riemann region, and cuts corresponding to different pairs meet at $\mathfrak{y}$ but do not cross. Hence the p pairs of crosscuts U_l, V_l make $\mathfrak{R}$ into a simply connected surface $\mathfrak{F}$ bounded in the positive sense by the curve

$$ T = U_1 V_1 U_1^{-1} V_1^{-1} \cdots U_p V_p U_p^{-1} V_p^{-1} $$

starting at the point $\mathfrak{y}$. Our construction maps the surface $\mathfrak{F}$ topologically onto the polygon $\mathfrak{R}_p$. Incidentally, our discussion makes it clear that the cuts U_l, V_l can be chosen to be piecewise smooth curves or even polygons whose sides are arcs of great circles on $\mathfrak{R}$. Furthermore, deformation of the system of cuts shows that the position of the point $\mathfrak{y}$ is of no consequence. The p pairs of crosscuts yield a canonical dissection of $\mathfrak{R}$ and $\mathfrak{R}_p$.

3. The fundamental group

We shall now construct the fundamental group Γ of a compact Riemann region $\mathfrak{R}$. In view of the definition in Section 8 of Chapter 1 this amounts to splitting the closed curves issuing from some fixed point $\mathfrak{z}_0$ on $\mathfrak{R}$ into homotopy classes. Two such curves are called *homotopic*, if it is possible to deform one into the other continuously with $\mathfrak{z}_0$ staying fixed. The definition of the product $C_1 C_2$ of two curves C_1 and C_2 issuing from $\mathfrak{z}_0$ carries over to the homotopy classes, and it is the latter which form the fundamental group. It is clear that, as an abstract group, the fundamental group is not affected by topological mappings, and therefore it suffices to determine the fundamental groups of the normal regions $\mathfrak{R}_p$.

It is not apparent that the Riemann region $\mathfrak{R}$ determines the number p uniquely; after all, in constructing the polygon $\mathfrak{N}_p$ we made use of auxiliary curves which could in general be selected in more than one way. In the next section, however, we shall show that the value of p is the same for all admissible constructions. What is more, we shall show using a group-theoretical argument, that for $p \neq q$ the fundamental groups of $\mathfrak{R}_p$ and $\mathfrak{R}_q$ are essentially different, that is, not isomorphic.

For $p = 0$, $\mathfrak{R}_p$ is a schlicht sphere. On such a surface every closed curve is homotopic to zero, so that, in this case, the fundamental group consists of the identity ε alone. Now let $p > 0$. For $p = 1$, $\mathfrak{R}_p$ is a torus and its fundamental group was determined in Section 9 of Chapter 1; this group will come up again in the following investigation. As in the preceding section, we denote by G_l, H_l ($l = 1, \ldots, p$) the p pairs of crosscuts in a canonical dissection of the surface issuing from a common point $\mathfrak{a}$, and by A_l, A_l' and B_l, B_l', the corresponding pairs of sides of the regular polygon $\mathfrak{N}_p = \mathfrak{N}$. Let a_l, a_l' and b_l, b_l' be the centers of these sides, c the center of the polygon, and $\mathfrak{a}_l = \mathfrak{a}_l'$, $\mathfrak{b}_l = \mathfrak{b}_l'$, $\mathfrak{c}$, the corresponding points on $\mathfrak{R}_p$. To the point $\mathfrak{a}$ there correspond all $4p$ vertices of the polygon; in particular, we denote the vertex formed by the sides B_p' and A_1 by a. In determining the fundamental group we may restrict ourselves to piecewise smooth curves which have finitely many points in common with the cuts G_l, H_l. The image on $\mathfrak{R}_p$ of the segments from c to a_l and from a_l' back to c is a closed path going from $\mathfrak{c}$ through $\mathfrak{a}_l$ and back to $\mathfrak{c}$ which intersects only the cut G_l, and that from left to right. We call this path S_l ($l = 1, \ldots, p$). The path T_l is defined somewhat differently, as the image of the segments from c to b_l' and from b_l back to c; as such it goes from $\mathfrak{c}$ through $\mathfrak{b}_l$ and back to $\mathfrak{c}$ and intersects H_l from right to left (Figure 47). Now let σ_l and τ_l be the homotopy classes of S_l and T_l. We shall prove that these $2p$ classes generate the full fundamental group Γ.

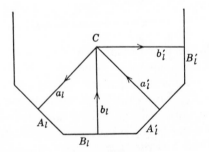

Figure 47

To this end we consider an arbitrary closed curve W on $\Re_p$ issuing from $\mathfrak{c}$. In determining the homotopy class of W we may assume that it does not pass through the point $\mathfrak{a}$. W meets the crosscuts at a finite number of points, all of which may be assumed to lie among the points $\mathfrak{a}_l$, $\mathfrak{b}_l$, and it may further be assumed that W crosses each crosscut that it intersects; this state of affairs can always be attained by means of a suitable deformation. Let $\mathfrak{c}_1, \ldots, \mathfrak{c}_r$ denote the successive points of intersection of W with the crosscuts. Then there is a continuous deformation which leaves $\mathfrak{c}_k$ and $\mathfrak{c}_{k+1}$ fixed and is such that the subarc of W between $\mathfrak{c}_k$ and $\mathfrak{c}_{k+1}$ passes through $\mathfrak{c}$ without intersecting a cut. In that case W consists of r successive arcs W_k ($k = 1, \ldots, r$) going from $\mathfrak{c}$ through $\mathfrak{c}_k$ and back to $\mathfrak{c}$ (provided, of course, $r > 0$). To the two subarcs of W_k determined by $\mathfrak{c}_k$ there correspond on $\Re$ two curves which go from c to a midpoint of a side and from the associated midpoint back to c. In view of the simply connected nature of $\Re$ this implies that W_k is homotopic to one of the $4p$ paths S_l, T_l, S_l^{-1}, T_l^{-1} ($l = 1, \ldots, p$). Hence the homotopy class of W is a product of r factors of the form σ_l, τ_l, σ_l^{-1}, τ_l^{-1} and, in the case $r = 0$, W is homotopic to zero. This completes the proof of our assertion.

If α and β belong to a group, then we put

$$\alpha\beta\alpha^{-1}\beta^{-1} = [\alpha, \beta],$$

and call this product the commutator of α and β.

Theorem 1: The generators σ_l, τ_l ($l = 1, \ldots, p$) of the fundamental group of $\Re_p$ satisfy the relation

$$\prod_{l=1}^{p} [\sigma_l, \tau_l] = \varepsilon.$$

Proof: We must show that the path

(1) $$W = S_1 T_1 S_1^{-1} T_1^{-1} \cdots S_p T_p S_p^{-1} T_p^{-1}$$

on $\Re_p$ is homotopic to zero. To this end we mark off on each side of the polygon $\Re$ the two points which are at a distance r away from its end points, where r is supposed less than half the side length. Let $e_l, f_l, g_l, h_l, f_l', e_l', h_l', g_l'$ be the new points, in this order, on the sides A_l, B_l, A_l', B_l' traversed in the positive sense. We deform the path $S_l T_l S_l^{-1} T_l^{-1}$ by moving the subsegments ca_l, $a_l'c$; cb_l', $b_l c$; ca_l', $a_l c$; cb_l, $b_l'c$ associated with it to ce_l, $e_l'c$; ch_l', $h_l c$; cf_l', $f_l c$; cg_l, $g_l'c$. We observe that then the image of W on $\Re$ is homotopic to the polygonal curve made up of ce_1; $e_l'h_l'$, $h_l f_l'$, $f_l g_l$, $g_l'e_{l+1}$ ($l = 1, \ldots$, $p - 1$); $e_p'h_p'$; $h_p f_p'$, $f_p g_p$, $g_p'c$ (Figure 48). As $r \to 0$ all of these segments, with the exception of the first and last, shrink to the $4p - 1$ vertices of the polygon distinct from a, and ce_1 and $g_p'c$ go over into ca and ac. If L is the path on $\Re_p$ corresponding to ca, then L joins $\mathfrak{c}$ and $\mathfrak{a}$ and, since $LL^{-1} \sim 0$, it follows that $W \sim 0$, which is what we wished to prove.

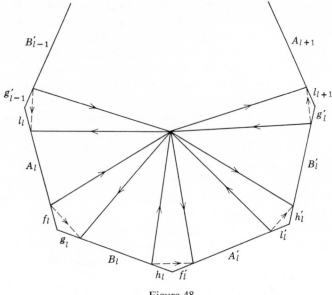

Figure 48

We shall show in the sequel that, in a sense which is yet to be made precise, Theorem 1 gives all the relations which hold for the generators of the fundamental group. As a preliminary step we introduce a construction which, for $p = 1$, reduces to the covering of the plane with a net of squares and, for arbitrary values of p is the analogue of this covering. We prove first that the path defined by (1) also yields a canonical dissection of $\Re_p$ with a replaced by c. S_l is represented on $\Re$ by the segments ca_l and $a'_l c$, which we now denote as C_{l1} and C_{l2}. Similarly, D_{l1} and D_{l2} denote the segments cb'_l and $b_l c$ which represent T_l. Furthermore, let A_{l1}, A_{l2} and B_{l1}, B_{l2} be the two pairs of halves into which a_l and b_l divide A_l and B_l. The corresponding halves of A'_l and B'_l are A'_{l2}, A'_{l1} and B'_{l2}, B'_{l1}. Now we cut $\Re$ along the $4p$ segments C_{l1}, C_{l2}, D_{l1}, D_{l2} into $4p$ congruent sectors, each of which is a quadrilateral with two opposite right angles. Next we transform each of these quadrilaterals into a congruent quadrilateral so that for $p > 1$ the obtuse angle replaces the acute angle. Although this transformation interchanges the lengths of the sides we leave their designations unchanged. The dissection just carried out introduces the associated edges C'_{l1}, C'_{l2}, D'_{l1}, D'_{l2}. By joining A_{l1}, A_{l2}, B_{l1}, B_{l2} to A'_{l1}, A'_{l2}, B'_{l1}, B'_{l2} we can construct out of the resulting $4p$ quadrilaterals a regular $4p$-gon $\Re^*$ in the manner made clear by Figure 49. That our joining prescription actually yields a polygon follows from the cyclic arrangement of the indices about the point a. It is clear that $\Re$ and $\Re^*$ are mirror images under the reflection which interchanges A with

C and B with D. To the path W there corresponds the boundary of $\mathfrak{N}^*$ traversed in the negative sense. In view of the simple-connectedness of $\mathfrak{N}^*$, this yields another proof of Theorem 1. We call $\mathfrak{N}^*$ the *dual* of $\mathfrak{N}$. It is easy to see that $\mathfrak{N}$ is also the dual of $\mathfrak{N}^*$, though we shall not make use of this in the sequel.

We now wish to imitate the method of construction of $\mathfrak{N}^*$; but this time, in place of the sectors, we propose to join about the vertex a, in the same schlicht manner, full copies of the polygon $\mathfrak{N}$. In euclidean geometry this can be done without deformations only in the case $p = 1$. In noneuclidean geometry for every $p > 1$ there is a regular polygon with $4p$ sides and angle sum 2π. The connection with noneuclidean geometry, however, will not be taken up until the third chapter (Vol. II). In the meantime we shall transform the polygons to be joined into polygons so narrow that they do not overlap when joined, and so that the angles which meet at the vertex add up to 2π. Also, we can require that the transformed polygons are convex, and that the images of C_{l1}, C_{l2}, D_{l1}, D_{l2} $(l = 1, \ldots, p)$ in them are again segments connecting the center of the polygon with the midpoints of the sides. The manner of joining the polygons is made clear in Fig. 50. There $\mathfrak{N}$ is hatched and the boundary of $\mathfrak{N}^*$ is marked by a broken line. Finally, we take countably many copies of the polygon $\mathfrak{N}$ and propose to make them into a schlicht, simply-connected surface in the plane by proper transformations and identifications. Let $\mathfrak{P}_0$, $\mathfrak{P}_1$, . . . denote the convex polygons in question. We bear in mind that to obtain a schlicht covering of the plane by means of squares, we can start out with a single square which we surround by a succession of wreaths of $8n$ squares, $n = 1, 2, \ldots$. In our case we start out with the polygon $\mathfrak{N} = \mathfrak{P}_0$. Next we attach externally to each of its $4p$ sides A_l, A_l', B_l, B_l' a properly transformed polygon along the corresponding sides A_l', A_l, B_l', B_l. and then at each of the $4p$ vertices of $\mathfrak{N}$ we adjoin $4p - 3$ additional polygons as described above. In this way we add to $\mathfrak{P}_0$, $4p (4p - 2)$ polygons $\mathfrak{P}_1$, $\mathfrak{P}_2$, . . . which form the first wreath. These polygons and the

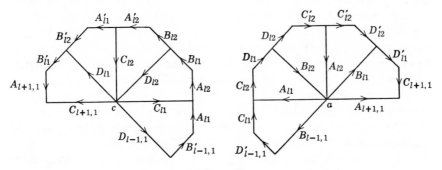

Figure 49

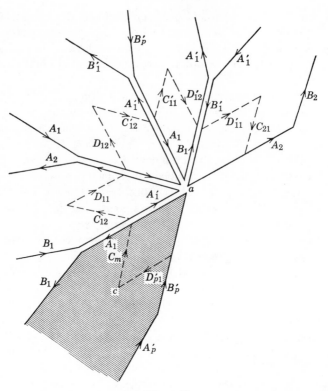

Figure 50

polygon $\mathfrak{P}_0$ form a surface $\mathfrak{B}_1$ bounded by a simple polygonal curve, and the number of the subpolygons $\mathfrak{P}_k$ which meet at any boundary point never exceeds two. Next, observing correspondence, we attach externally to each boundary edge of $\mathfrak{B}_1$ a properly transformed copy of $\mathfrak{N}$ and then fill in the gaps at the vertices lying on the boundary. The new polygons $\mathfrak{P}_k$ form the second wreath and, together with $\mathfrak{B}_1$, a larger surface $\mathfrak{B}_2$. If we see to it that the polygons in the second wreath do not overlap, then $\mathfrak{B}_2$ is bounded by a simple polygonal curve, and the number of subpolygons which meet at any boundary point never exceeds two. Since at each step the number of boundary edges increases, this procedure can be continued indefinitely and gives rise to an increasing sequence of simply-connected schlicht regions $\mathfrak{B}_1$, $\mathfrak{B}_2$, ... which converge to a simply-connected region $\mathfrak{B}$. We need not concern ourselves with the boundary of $\mathfrak{B}$. Each of the polygons $\mathfrak{P}_k$ has a dual $\mathfrak{P}_k^*$ which stands in the same relation to $\mathfrak{P}_k$ as $\mathfrak{N}^*$ to $\mathfrak{N}$. Just as the polygons $\mathfrak{P}_0$, $\mathfrak{P}_1$, ... so, too, the polygons $\mathfrak{P}_0^*$, $\mathfrak{P}_1^*$, ... yield a simple covering of the region $\mathfrak{B}$ free of gaps. After these preliminaries we are ready to formulate the next theorem whose statement will become clearer and more precise in the process of proof.

Theorem 2: The fundamental group of $\mathfrak{R}_p$ is the group with $2p$ generators σ_l, τ_l $(l = 1, \ldots p)$ and the single defining relation

$$\prod_{l=1}^{p} [\sigma_l, \tau_l] = \varepsilon.$$

Proof: Let W again denote a closed curve on $\mathfrak{R}_p$ issuing from $\mathfrak{c}$ which intersects the $2p$ crosscuts G_l, H_l $(l = 1, \ldots, p)$ only in a finite number of points. We wish to transplant them continuously to $\mathfrak{B}$ and proceed in a manner analogous to that of Section 9 in Chapter 1, in which we treated the case corresponding to $p = 1$. To this end we follow the image curve of W on the polygon $\mathfrak{P}_0 = \mathfrak{R}$ beginning with the initial point c, except that when we encounter a boundary point of $\mathfrak{P}_0$ we do not shift to the corresponding boundary edge of $\mathfrak{P}_0$, but rather enter the appropriate polygon which is joined to $\mathfrak{P}_0$ on $\mathfrak{B}$. We note that we may now admit the possibility that W passes through $\mathfrak{a}$, for, as a result of appropriate identification, we managed to construct at all vertices of the polygons complete neighborhoods which are images of the neighborhoods of $\mathfrak{a}$ on $\mathfrak{R}_p$. It is clear that this argument can be repeated with any polygon $\mathfrak{P}_k$ in place of $\mathfrak{P}_0$. In this way we obtain on $\mathfrak{B}$ a unique image curve $\tilde{W}$ of W issuing from the point c on $\mathfrak{P}_0$. Since W is closed on $\mathfrak{R}_p$, that is, it returns to $\mathfrak{c}$, it follows that the end point of $\tilde{W}$ is a point of a polygon $\mathfrak{P}_k$ which corresponds to the point $\mathfrak{c}$, that is, the center of gravity. Conversely, a curve on $\mathfrak{B}$ which joins the center of $\mathfrak{R}$ to the center of gravity of one of the polygons $\mathfrak{P}_k$ is the image of a closed curve on $\mathfrak{R}_p$ issuing from $\mathfrak{c}$.

A continuous deformation of W which leaves $\mathfrak{c}$ fixed induces a continuous deformation of $\tilde{W}$ which leaves the two end points fixed, and conversely. If W on $\mathfrak{R}_p$ is homotopic to zero then $\tilde{W}$ on $\mathfrak{B}$ must be closed, that is it must return to c. The converse of the latter assertion is also true since $\mathfrak{B}$ is simply connected, a fact of which we make essential use at this junction. From now on it suffices to assume that W is a product of the arbitrarily repeated factors $S_l, S_l^{-1}, T_l, T_l^{-1}$, for these products already yield representatives of all the homotopy classes. We denote the corresponding polygonal curve $\tilde{W}$ on $\mathfrak{B}$ by means of the analogous product in the factors $\sigma_l, \sigma_l^{-1}, \tau_l, \tau_l^{-1}$. For the moment then, the symbols σ_l, τ_l denote the images of S_l, T_l on $\mathfrak{B}$ rather then homotopy classes. If we form all the words in the symbols $\sigma_l, \sigma_l^{-1}, \tau_l, \tau_l^{-1}$ $(l = 1, \ldots, p)$ barring cyclic variants, then we obtain all the polygonal paths $\tilde{W}$ to be considered, each appearing exactly once. Our aim now is to determine for which words the path $\tilde{W}$ is closed.

From now on let $\tilde{W}$ be closed. To begin with, we consider only the special case when $\tilde{W}$ is a simple closed curve, and therefore the complete boundary of a polygonal surface $\mathfrak{Q}$ on $\mathfrak{B}$. The example closest at hand is

$$\tilde{W} = \prod_{l=1}^{p} S_l T_l S_l^{-1} T_l^{-1};$$

then $\tilde{W}$ is just the negatively traversed boundary of the polygon $\mathfrak{P}_0^*$ on $\mathfrak{B}$ dual to $\mathfrak{N} = \mathfrak{P}_0$ and is represented by the word

$$\kappa = \prod_{l=1}^{p} [\sigma_l, \tau_l].$$

If ν is a word in σ_l, τ_l whose corresponding path on $\mathfrak{B}$ joins c to the center of gravity of the polygon $\mathfrak{P}_k$, then, clearly $\nu\kappa\nu^{-1}$ yields a loop issuing from c which traverses the boundary of the dual polygon $\mathfrak{P}_k^*$ in the negative sense. Since the dual polygons provide a covering of $\mathfrak{B}$ and the boundary $\tilde{W}$ of $\mathfrak{Q}$ consists of sides of these polygons, it follows that $\mathfrak{Q}$ is covered simply and without gaps by a finite number of these polygons. If $\mathfrak{Q}$ consists of just one such polygon then this polygon is $\mathfrak{P}_0^*$, and in this case $\tilde{W}$ is represented by κ or κ^{-1}. Otherwise, we show that among the polygons $\mathfrak{P}_k^*$ which cover $\mathfrak{Q}$ and touch its boundary there is at least one which can be removed leaving the remaining polygonal surface simply connected. It is clear that $\mathfrak{P}_k^*$ has the required property if and only if this polygon touches the boundary $\tilde{W}$ of $\mathfrak{Q}$ along a connected polygonal curve. If this is not the case, let L denote a subpolygonal curve on $\tilde{W}$ both of whose end points belong to $\mathfrak{P}_k^*$ (Figure 51). Then the covering polygons $\mathfrak{P}_l^*$ which touch L cannot touch $\tilde{W}$ except on L. If $\mathfrak{P}_l^*$ also fails to have the required property, then we can once more employ the reasoning used for $\mathfrak{P}_k^*$ and reach our aim after a finite number of steps, because the parts of the boundary $L, \ldots$ become progressively smaller. Our argument also shows that it is even possible to find a polygon $\mathfrak{P}_k^*$ of the required type which is different from $\mathfrak{P}_0^*$. Let the boundary $\tilde{W}$ of $\mathfrak{Q}$ be negatively oriented. Then the word

$$\omega = \alpha\beta\gamma$$

which represents $\tilde{W}$ consists of three subwords with β corresponding to the part of $\tilde{W}$ on $\mathfrak{P}_k^*$. One of the $4p$ cyclic variants of κ begins with β, that is, has the form $\beta\delta$. On the other hand, this variant arises from κ if, for a certain decomposition of $\kappa = \lambda\mu$ of κ into two subwords, we write $\mu\lambda$. It follows that the word β arises from the word $\lambda^{-1}\kappa\lambda\delta^{-1}$ if we replace κ by $\lambda\mu$, delete $\lambda^{-1}\lambda$, then replace $\mu\lambda$ by $\beta\delta$ and delete $\delta\delta^{-1}$. We make use of the terminology in Section 2 and speak of reduction of a word when we delete from it subwords of the form $\rho\rho^{-1}$ or $\rho^{-1}\rho$. Insertion of such a subword will be referred to as extension of the word. The congruence

$$\omega \equiv \theta$$

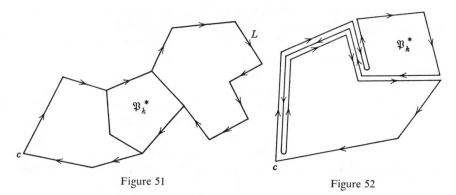

Figure 51   Figure 52

is to signify that the two words ω and θ are the same apart from reductions and extensions. In particular,

$$\beta \equiv \lambda^{-1}\kappa\lambda\delta^{-1}, \qquad \omega \equiv (\alpha\lambda^{-1})\kappa(\alpha\lambda^{-1})^{-1}(\alpha\delta^{-1}\gamma).$$

Here the polygonal curve represented by the word $\alpha\delta^{-1}\gamma$ is just the negatively traversed boundary of the polygon obtained from $\mathfrak{Q}$ by dropping $\mathfrak{P}_k^*$ (Figure 52). Using induction on the number h of dual polygons covering $\mathfrak{Q}$ we obtain

$$(1) \qquad \omega \equiv \prod_{l=1}^{h} (\nu_l \kappa \nu_l^{-1}),$$

where $\nu_1, \ldots, \nu_h$ are certain words. This formula remains valid when $\tilde{W}$ is positively oriented, provided that κ is replaced by κ^{-1}.

It remains to consider the case when the closed path $\tilde{W}$ contains double points. By possibly going over to a congruent path we may suppose the word ω representing $\tilde{W}$ reduced. Using a double point on $\tilde{W}$ we can decompose $\tilde{W}$ into three ordered parts of which the middle part is a closed polygonal curve, and the last part is perhaps the single point c. The corresponding decompositions into subwords are $\omega = \alpha\beta\gamma$ and $\omega = \alpha\beta$ with β representing a closed path on $\mathfrak{B}$. Since the words $\alpha\gamma$ and β or α and β contain fewer symbols than ω, we can apply an induction argument on the number of symbols in ω and make use of the relation $\omega \equiv (\alpha\gamma)(\gamma^{-1}\beta\gamma)$ (Figure 53). Using (1) we obtain quite generally

$$(2) \qquad \omega \equiv \prod_{l=1}^{h} (\nu_l \kappa \nu_l^{-1})^{p_l},$$

for an appropriate value of h and for $p_l = 1$ or -1. Conversely, the meaning of the symbols on the right side of (2) implies readily that the path $\tilde{W}$ represented by ω in (2) is closed.

Now let σ_l, τ_l $(l = 1, \ldots, p)$ again be the generators of the fundamental group Γ of $\mathfrak{R}_p$. A relation in these generators can be put in the form $\omega = \varepsilon$, where ω is a word in the symbols σ_l, τ_l, σ_l^{-1}, τ_l^{-1}. If we replace the symbols in this relation by S_l, T_l, S_l^{-1}, T_l^{-1}, then we obtain a closed path W on $\mathfrak{R}_p$ issuing

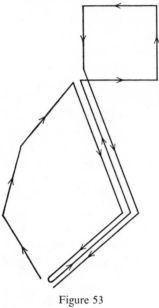

from $\mathfrak{c}$. The relation asserts that W is homotopic to zero. But, in view of (2), it is possible to change the word ω by means of reductions and extensions to the word on the right side of (2), and then the special relation $\kappa = \varepsilon$ proved in Theorem 1 has as an immediate consequence the relation $\omega = \varepsilon$. We note that if we introduce the free group $\bar{\Gamma}$ on the symbols σ_l, τ_l $(l = 1, \ldots, p)$, then formula (2) holds with equality in place of equivalence, for in a free group the elements are similarly unaffected by reductions or extensions. If ν varies over the elements of $\bar{\Gamma}$, then the conjugates of κ are given by $\nu \kappa \nu^{-1}$ and all these conjugates generate a subgroup Δ of $\bar{\Gamma}$. It is obvious that Δ is a normal subgroup of $\bar{\Gamma}$. Two elements α and β of $\bar{\Gamma}$ belong to the same coset of Δ if $\beta \alpha^{-1} = \omega$ is in Δ, and these cosets form the factor group $\bar{\Gamma}/\Delta$. If we again think of the elements of $\bar{\Gamma}$ as homotopy classes, then they form precisely the group Γ, and $\alpha = \beta$ if and only if $\omega = \varepsilon$. We showed that such an element ω of $\bar{\Gamma}$ is in Δ. Hence

Figure 53

(3) $$\Gamma = \bar{\Gamma}/\Delta.$$

In particular, this proves Theorem 2 and yields a more precise formulation of this theorem.

It should be noted that the proof just presented gives a constructive method for deciding whether two given paths on $\mathfrak{R}_p$ are homotopic or not. The procedure for finding the right-hand side of (2) is also constructive in nature. In the case $p = 1$ the fundamental group is commutative, for in that case the two generators $\sigma = \sigma_1$, $\tau = \tau_1$ are, by Theorem 1, connected by the relation $\sigma \tau \sigma^{-1} \tau^{-1} = \varepsilon$, and this implies that $\sigma \tau = \tau \sigma$. For $p > 1$ the fundamental group is not commutative, for then the path $C_1 D_1 C_1' D_1'$ is only part of the boundary of $\mathfrak{R}^* = \mathfrak{P}_0^*$ and is, therefore, not closed on $\mathfrak{B}$. But then $\sigma_1 \tau_1 \sigma_1^{-1} \tau_1^{-1} \neq \varepsilon$.

Let $\tilde{W}$ be a curve on $\mathfrak{B}$ connecting the point $\mathfrak{c}$ to the center of gravity s of an arbitrarily chosen polygon $\mathfrak{P}_k$, and let ω be the homotopy class of that closed curve on $\mathfrak{R}_p$ issuing from $\mathfrak{c}$ whose image is $\tilde{W}$. Then ω depends on s alone so that we can denote the polygons $\mathfrak{P}_k$ in a unique manner by $\mathfrak{P}_\omega$. Now we make use of the fundamental group Γ to define the covering surface $\mathfrak{U}$ of the compact Riemann region $\mathfrak{R}$ as follows: Just as at the end of Section 2, let $\mathfrak{F}$ denote the Riemann region dissected canonically by means of p pairs of crosscuts U_l, V_l $(l = 1, \ldots, p)$ and let the point $\mathfrak{y}_0$ of $\mathfrak{F}$ correspond to the center c of $\mathfrak{R}$. To the path S_l in the homotopy class σ_l there corresponds on $\mathfrak{R}$ a closed curve issuing from $\mathfrak{y}_0$ which crosses the single cut U_l from left to right, and similar statements hold for T_l, τ_l and V_l. We now associate with every element ω of the fundamental group a copy $\mathfrak{F}_\omega$ of $\mathfrak{F}$. If $\omega_1^{-1}\omega_2$ or $\omega_2^{-1}\omega_1$ is equal to one of the generators σ_l, τ_l, then we call $\mathfrak{F}_{\omega_1}$ and $\mathfrak{F}_{\omega_2}$ neighbors and join them once in the appropriate manner along the cut U_l or V_l. This gives rise to a connected surface $\mathfrak{U}$, which is constructed out of the $\mathfrak{F}_\omega$ in just the same way as the schlicht plane surface $\mathfrak{B}$ is constructed out of the polygons $\mathfrak{P}_\omega$. It follows that $\mathfrak{U}$ is simply connected and that two paths on $\mathfrak{U}$ issuing from $\mathfrak{y}_0$ lead to the same point if and only if their traces on $\mathfrak{R}$ are homotopic. It is easy to infer from this property that the covering surface $\mathfrak{U}$ is independent of the canonical dissection of $\mathfrak{R}$ and is a Riemann region; in fact, here is how $\mathfrak{U}$ can be constructed out of circular regions. If $\mathfrak{G}$ is a disk on $\mathfrak{R}$, then we associate with every element of the fundamental group a copy $\mathfrak{G}_\lambda$ of $\mathfrak{G}$ which is to lie on $\mathfrak{U}$ over $\mathfrak{G}$. If $\mathfrak{H}$ is another disk on $\mathfrak{R}$ and $\mathfrak{H}_\mu$ is the disk on $\mathfrak{U}$ associated with an arbitrary μ in the fundamental group, then we must give a rule when $\mathfrak{G}_\lambda$ and $\mathfrak{H}_\mu$ are to be joined. Of course, we need only consider the case when $\mathfrak{G}$ and $\mathfrak{H}$ on $\mathfrak{R}$ have a nonempty intersection. We choose on $\mathfrak{R}$, independently of λ, a path G going from $\mathfrak{y}_0$ to the center $\mathfrak{g}$ of $\mathfrak{G}$ and, furthermore, a closed path L issuing from $\mathfrak{y}_0$ and belonging to the homotopy class λ. We define similarly $\mathfrak{h}$, H, M for $\mathfrak{H}$. Further, let $D_{\mathfrak{g}\mathfrak{h}}$ be a path on $\mathfrak{R}$ from $\mathfrak{g}$ to $\mathfrak{h}$ belonging to $\mathfrak{G} \cup \mathfrak{H}$. We stipulate that $\mathfrak{G}_\lambda$ and $\mathfrak{H}_\mu$ are to be joined if and only if the path $LGD_{\mathfrak{g}\mathfrak{h}} H^{-1}M^{-1}$ on $\mathfrak{R}$ is homotopic to zero. For this definition to make sense we must show that

$$\mathfrak{G}_\lambda \cap \mathfrak{H}_\mu \cap \mathfrak{I}_\nu = (\mathfrak{G}_\lambda \cap \mathfrak{H}_\mu) \cap (\mathfrak{H}_\mu \cap \mathfrak{I}_\nu).$$

If $\mathfrak{G} \cap \mathfrak{H} \cap \mathfrak{I}$ is not empty and $\mathfrak{i}$ is the center of $\mathfrak{I}$ then

$$D_{\mathfrak{g}\mathfrak{h}} D_{\mathfrak{h}\mathfrak{i}} \sim D_{\mathfrak{g}\mathfrak{i}},$$

and the relations

$$LGD_{\mathfrak{g}\mathfrak{h}}H^{-1}M^{-1} \sim E, \qquad MHD_{\mathfrak{h}\mathfrak{i}}J^{-1}N^{-1} \sim E$$

imply

$$LGD_{\mathfrak{g}\mathfrak{i}}J^{-1}N^{-1} \sim E.$$

In order to obtain in this way the branch points of $\mathfrak{U}$ we must treat the surface elements at the branch points of $\mathfrak{R}$ in the same way as we treated the disks $\mathfrak{G}$.

It should be pointed out that there is a much more general notion of a covering surface $\mathfrak{T}$ of a Riemann region $\mathfrak{R}$ than the one above. This more general notion arises if we require only that every analytic function on $\Re$ is always analytic on $\Im$. Such general covering surfaces need not be simply connected and may exhibit new branch points. In particular, every Riemann region is, in this sense, a covering surface of the number sphere. The earlier covering surface $\mathfrak{U}$ is then identified as the universal covering surface of $\Re$. In the sequel we shall deal only with covering surfaces in the earlier sense of the term. We note also that the covering surface of the Riemann surface $\Re_4$ discussed in Sections 8 and 10 of Chapter 1 fits our present discussion in the special case $p = 1$.

One final remark. The free group Γ with the $2p$ generators σ_l, τ_l ($l = 1, \ldots, p$), which appeared above, has simple topological significance. This group is the homotopy group of the surface $\dot{\Re}_p$ obtained by omitting the point $\mathfrak{a}$ from $\Re_p$. The proof of this assertion follows from arguments similar to those employed in proving Theorem 2, except that in the present case these arguments are frequently simpler. We shall not pursue this matter since it is of no relevance in the sequel.

4. Invariance of the genus

We obtained the sphere with p handles out of the compact Riemann region in the following manner. There were the distinct points $z_1, \ldots, z_m$ of the schlicht number sphere over which lay the branch points of $\Re$ and an additional point z^* distinct from $z_1, \ldots, z_m$. Then, by means of m cuts $L_1, \ldots, L_m$ from z^* to z_j ($j = 1, \ldots, m$), we changed the sphere into the surface $\mathfrak{S}$ with the n sheets $\mathfrak{S}_1, \ldots, \mathfrak{S}_n$ of $\Re$ lying over it. These sheets were mapped topologically onto polygons $\mathfrak{P}_1, \ldots, \mathfrak{P}_n$, which were suitably put together to form a single polygon $\mathfrak{P}$. Finally, by means of certain changes $\mathfrak{P}$ was made into a regular $4p$-gon $\Re$ with ordered sides A_k, B_k, A_k', B_k' ($k = 1, \ldots, p$) which, after identification of corresponding sides, became the sphere with p handles. It should be pointed out that the construction of $\mathfrak{P}$ and $\Re$ was definitely not the result of a uniquely determined procedure, and hence it is not clear to what extent the value of p depends on the manner of construction of the polygon $\Re$.

Theorem 1: The number p is uniquely determined by the compact Riemann region $\Re$.

Proof: If we think of the cut curves $L_1, \ldots, L_m$ as marked on $\Re$, then we can view $\Re$ as a polyhedron whose faces are the n sheets $\mathfrak{S}_1, \ldots, \mathfrak{S}_n$, whose edges are the cuts, and whose vertices are the points of the Riemann region lying over the $m + 1$ traces $z^*, z_1, \ldots, z_m$. If E is the number of vertices, K the number of edges, and F the number of faces of our polyhedron, then $F = n$ and $K = mn$; the latter relation follows from the fact that each of the m cuts $L_1, \ldots, L_m$ yields two cuts per sheet and these are joined in pairs on $\Re$. To determine the value of E we must determine the number of points of the Riemann region $\Re$ lying over each of the points $z^*, z_1, \ldots, z_m$. There are exactly n ordinary points of $\Re$ over z^*. Let r_j be the number of branch points and ordinary points over z_j $(j = 1, \ldots, m)$, and let $l_j^{(1)}$, $l_j^{(2)}, \ldots, l_j^{(r_j)}$ be the associated numbers of connected sheets. We call the sum

$$(l_j^{(1)} - 1) + (l_j^{(2)} - 1) + \cdots + (l_j^{(r_j)} - 1) = n - r_j$$

the *branch number of the point* z_j, and the sum

$$v = \sum_{j=1}^{m} (n - r_j) = mn - (r_1 + \cdots + r_m)$$

the *branch number of the Riemann region*. If there are no branch points over z_j, then the value of its branch number is $n - r_j = 0$. Hence v depends on $\Re$ alone. Further, $v \geq 0$ and $v = 0$ only if there exist no branch points, that is, when $n = 1$. Using v we obtain the following formula for the number of vertices of our polyhedron:

$$E = n + (r_1 + \cdots + r_m) = n + mn - v.$$

Now we form the expression

(1) $$E - K + F = 2n - v,$$

called the *Euler characteristic* of the polyhedron. The right-hand side of (1) depends on $\Re$ alone and not on the manner of dissection. Now we show that, on the other hand, the value of the left-hand side of (1) remains unchanged when we go over to $\mathfrak{P}$ and $\mathfrak{N}$, provided that we make proper identifications of sides and vertices. The n sheets of $\Re$ were mapped onto the polygons $\mathfrak{P}_1, \ldots, \mathfrak{P}_n$ with total number of $2mn$ sides. These are to be identified in pairs, and we count the two sides of such a pair as a single edge in accordance with the definition of an edge of the original polyhedron. Similarly, all polygonal vertices to which there corresponds on $\Re$ a single point are counted as a single vertex. In the course of construction of $\mathfrak{P}$, whenever two polygons are joined two sides of a pair, that is an edge, are lost. At the same time the number F of available faces decreases by one while E remains unchanged. It follows that transition to $\mathfrak{P}$ leaves the characteristic unchanged, but then for $\mathfrak{P}$ we have $F = 1$.

In going from $\mathfrak{P}$ to $\mathfrak{N}$ we must consider the changes called reductions and diagonalizations. A reduction involved the joining of neighboring sides A, A' on the boundary of the polygon. This means that we still have $F = 1$, but the number K of edges decreases by one. Since the joining takes place from the outside, the interior angle between A and A' goes over into a round angle, and a full neighborhood of the corresponding vertex lies in the interior of the polygon arising from the reduction (Figure 54). Therefore no vertex of the polygon other than a can have the same image on $\mathfrak{R}$. Since, as a result of the joining, the vertex a disappears, it follows that the number of vertices decreases by one and the characteristic remains unchanged.

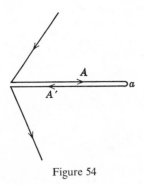

Figure 54

Diagonalization involved cutting the polygon along a diagonal D and joining the two resulting parts along two corresponding sides A, A'. This leaves $F = 1$ faces. The number K of edges is also unchanged for in place of the lost pair of sides A, A' arises the new pair of sides D, D'. Since the end points of D are polygonal vertices, the addition of D and D' does not increase the count of vertices and, since A and A' are not neighbors, the joining of A and A' does not decrease the count of vertices. Hence diagonalization does not affect the number E of vertices and the characteristic is again unchanged.

Now we compute the characteristic of $\mathfrak{R}$ or, what amounts to the same thing, the characteristic of the canonically dissected sphere with p handles. Here the edges are formed by the $2p$ crosscuts G_l, H_l ($l = 1, \ldots, p$), and their common point is the single vertex. Therefore, $K = 2p$, $E = 1$, and $F = 1$. Hence we have for $\mathfrak{R}_p$

(2) $$E - K + F = 2 - 2p,$$

and, in view of (1),

$$2 - 2p = 2n - v,$$

so that,

(3) $$p = \frac{v}{2} - n + 1.$$

This shows that the value of p depends on $\mathfrak{R}$ alone. This proves Theorem 1.

It follows directly from (3) that the branch number v is always even. If $\mathfrak{R}$ has two sheets, that is $n = 2$, then for any branch point that appears the local branch number has the value one, so that v coincides with the number of branch points. This is, therefore an even number $2h$ and, in view of (3), we have $p = h - 1$. In particular, for $2h = 2$, 4 we have $p = 0$, 1. This was anticipated by the work in Section 5 of Chapter 1 in which we mapped the two-sheeted Riemann surfaces with two and four branch points onto a sphere and a torus, respectively. If we consider quite generally a two-sheeted Riemann surface with $2p + 2$ arbitrarily given branch points, then we obtain examples with prescribed value of p; such examples were mentioned earlier. For $n = 1$, $\mathfrak{R}$ is the schlicht spherical surface $\mathfrak{R}$, and then $v = 0$, $p = 0$.

The number p is called the *genus* of the compact Riemann region $\mathfrak{R}$. In topology the genus is defined for arbitrary triangulable surfaces and is shown to be a topological invariant. To obtain this result one makes use of Poincaré's generalization of Euler's polyhedral formula which asserts that the numbers giving the vertices, edges, and faces of a two sided polyhedron with first Betti number $2p$ are connected by (2). We shall not go into these matters here. However, using the fundamental group we shall prove that for $p \neq q$ the surfaces $\mathfrak{R}_p$ and $\mathfrak{R}_q$ are not homeomorphic. Since we mapped $\mathfrak{R}$ topologically onto $\mathfrak{R}_p$, this implies, in particular, that two homeomorphic compact Riemann regions have the same genus.

Theorem 2: If $p \neq q$ then the fundamental groups of $\mathfrak{R}_p$ and $\mathfrak{R}_q$ are not isomorphic.

Proof: If α, β, γ are elements of a group Ω, then the commutator $[\alpha, \beta] = \alpha\beta\alpha^{-1}\beta^{-1}$ satisfies the relations $[\alpha, \beta]^{-1} = [\beta, \alpha]$, $\gamma[\alpha, \beta]\gamma^{-1} = [\gamma\alpha\gamma^{-1}, \gamma\beta\gamma^{-1}]$. It follows that all the finite products of commutators of elements of the given group Ω form a normal subgroup Λ called the commutator subgroup of Ω. In particular, let Ω be the free group on r generators $\sigma_1, \ldots, \sigma_r$. Every element α of Ω can be written uniquely as a reduced, ordered, finite power product in the generators. If in this representation of α the generator σ_k appears with successive exponents $l_1, \ldots, l_h$, then we call

$$l_1 + \cdots + l_h = e_k$$

the exponent sum of α associated with σ_k, and the r numbers $e_1, \ldots, e_n$, *the sum system of α.* The latter can be represented as a vector $\mathfrak{v}(\alpha)$ in r-dimensional space. It is clear that we have the relations

$$\mathfrak{v}(\alpha^{-1}) = -\mathfrak{v}(\alpha), \qquad \mathfrak{v}(\alpha\beta) = \mathfrak{v}(\alpha) + \mathfrak{v}(\beta),$$

so that the correspondence which sends α into $\mathfrak{v}(\alpha)$ is a homomorphic mapping of Ω into the additive group Φ of the vectors $\mathfrak{v}(\alpha)$. For any commutator each of the r exponent sums has the value 0, so that every element of the commutator group Λ is mapped to the zero vector $\mathfrak{v}(\varepsilon)$. On the other hand we have

$$\beta\alpha = \alpha\beta[\beta^{-1}, \alpha^{-1}],$$

and we can show by an induction argument based on this rule that every group element with sum system $e_1, \ldots, e_n$ can be put in the form

$$\alpha = \sigma_1^{e_1} \cdots \sigma_r^{e_r}\lambda,$$

where λ belongs to the commutator group. This implies that the factor group Ω/Λ is isomorphic to Φ. If we prefer a multiplicative group to the group Φ, then we can take instead of $\mathfrak{v}(\alpha)$ the power product $\sigma_1^{e_1} \cdots \sigma_r^{e_r}$.

But now the multiplication of the symbols $\sigma_1, \ldots, \sigma_r$ must be commutative. Then we see that the factor group of the free group on r generators with respect to its commutator subgroup is isomorphic to the free abelian group on r generators.

Now let $\Omega = \mathring{\Gamma}$ be the free group on $2p$ generators σ_l, τ_l $(l = 1, \ldots, p)$, Λ its commutator subgroup and, as before, Δ the subgroup generated by all conjugates $\alpha^{-1}\kappa\alpha$ with

$$\kappa = \prod_{l=1}^{p} [\sigma_l, \tau_l],$$

and α in $\mathring{\Gamma}$. Δ is a normal subgroup of $\mathring{\Gamma}$, and therefore also of Λ. In view of (3) in Section 3, the fundamental group Γ of $\mathfrak{R}_p$ is isomorphic to $\mathring{\Gamma}/\Delta$, and its commutator group is therefore isomorphic to Λ/Δ. The factor group of $\mathring{\Gamma}/\Delta$ relative to Λ/Δ is isomorphic to $\mathring{\Gamma}/\Lambda$, that is, to the free abelian group on $2p$ generators. The group-theoretical result just used could be avoided if we defined the sum system directly for the elements of the group Γ instead of introducing the free group Ω.

If $\mathfrak{R}_p$ and $\mathfrak{R}_q$ are homeomorphic then their fundamental groups are isomorphic, and the same is true for their commutator subgroups. But then the free abelian group on $2p$ generators is isomorphic to the free abelian group on $2q$ generators. If we had $p > q$ say, then, by considering linear equations we could show, by an argument similar to that used in the proof of Theorem 2 in Section 9 of Chapter 1, that $2q + 1$ generators of the former group must be dependent. This contradiction proves our theorem. The approach just used also proves the invariance of the genus of an arbitrary triangulable surface.

We return once more to the factor group of the fundamental group relative to its commutator subgroup. We obtain this group out of the fundamental group by making the latter commutative, that is, by replacing the single defining relation $\kappa = \varepsilon$ by the defining relations $\alpha\beta = \beta\alpha$, in which α and β vary independently over the $2p$ generators σ_l, τ_l $(l = 1, \ldots, p)$. The resulting free abelian group on $2p$ generators is called the *homology group* of the basic Riemann region $\mathfrak{R}$. Just as the homotopy group, so also the homology group was introduced by Poincaré in a more general context. The geometric significance of the homology group comes to the fore in Chapter 4 (Vol. II) in connection with the investigation of the periods of abelian integrals. Specifically, if C and D are two closed curves issuing from a point of $\mathfrak{R}$, then we have the following rule for the integration of single-valued functions:

$$\int_{CD} = \int_C + \int_D = \int_D + \int_C + \int_{DC}.$$

This rule states that the outcome is independent of the order of the two paths. On the other hand, in analytic continuation of a given function element, it makes in general a difference whether we continue along CD or DC, and here the homotopy group plays a decisive role.

5. The Poisson integral

In this and in the next three sections we wish to develop various auxiliary results which will enable us to treat in the last section of this chapter the following problem: Given a compact Riemann region $\mathfrak{R}$ of genus p, find an analytic function which maps the simply connected covering surface $\mathfrak{U}$ of $\mathfrak{R}$ conformally onto a schlicht domain $\mathfrak{F}$ in the complex plane or on the number sphere, where for $p = 0$ the domain is the complete number sphere, for $p = 1$ it is the complete complex plane, and for $p > 1$ it is the interior of the unit circle. In the special case when $\mathfrak{R}$ is two-sheeted and has exactly four branch points the solution is implied by Theorem 2 in Section 8 of Chapter 1. In that case $p = 1$ and the elliptic integral of the first kind yields the required solution. For $p > 1$ there correspond to the elliptic integral of the first kind the abelian integrals of the first kind of which there are p linearly independent ones. These integrals do not yield the required mapping, since the mapping effected by them is not schlicht in the plane.

The solution of the proposed problem requires lengthy preparations. In this connection the concept of the boundary value of a function, which we are about to introduce, will turn out to be of importance. Let $f(x, y)$ be a real or complex valued function defined in a schlicht domain $\mathfrak{G}$ of the (x, y)-plane and let $f(x, y)$ be continuous in $\mathfrak{G}$. We write, for brevity, $z = x + iy$ and $f(z)$ in place of $f(x, y)$; here $f(z)$ need not be an analytic function of the complex variable z. To begin with, the function is not defined at the boundary points of $\mathfrak{G}$. Let ζ be a boundary point of $\mathfrak{G}$. If for every sequence z_n ($n = 1$, $2, \ldots$) in $\mathfrak{G}$ which converges to ζ the corresponding sequence $f(z_n)$ of functional values converges to a limit, then this limit is independent of the special choice of the sequence $z_n \to \zeta$ and is called the *boundary value $f(\zeta)$* of the function $f(z)$ at the point ζ. If $f(z)$ has a boundary value at every point of $\mathfrak{G}$ then we say that it has boundary values throughout.

Theorem 1: Let f be a function which is continuous in $\mathfrak{G}$ and which has boundary values throughout. Let $\tilde{\mathfrak{G}}$ be the closure of $\mathfrak{G}$. If we assign to f at the boundary points its corresponding boundary values, then the resulting function f is continuous in all of $\tilde{\mathfrak{G}}$.

Proof: We need prove the continuity of f only at the boundary points ζ of $\mathfrak{G}$. Let $z_1, z_2, \ldots$ be a sequence of points in $\tilde{\mathfrak{G}}$ converging to ζ. We must prove that

$$(1) \qquad\qquad f(z_n) \to f(\zeta) \qquad (n \to \infty).$$

This convergence is assumed in the case when the z_n belong to the interior of $\tilde{\mathfrak{G}}$, and we can therefore suppose all of the z_n to be boundary points. If (1) were false for a certain sequence of boundary points, it would be possible to choose a positive number δ and a subsequence $\zeta_1, \zeta_2, \ldots$ such that

$$|f(\zeta_n) - f(\zeta)| > \delta \qquad (n = 1, 2, \ldots).$$

Since with each of the ζ_n there is associated a boundary value of the function, we can associate with each of the ζ_n an interior point η_n such that

$$|f(\zeta_n) - f(\eta_n)| < \frac{\delta}{2} \qquad (n = 1, 2, \ldots), \quad \eta_n \to \zeta \quad (n \to \infty).$$

From the last two inequalities we obtain by subtraction the inequality

$$|f(\eta_n) - f(\zeta)| > \frac{\delta}{2} \qquad (n = 1, 2, \ldots),$$

which contradicts the existence of the boundary value $f(\zeta)$ at ζ. This proves (1).

We shall now investigate the boundary values of harmonic functions. We recall that a real valued function $u = u(x, y) = u(z)$ which is twice continuously differentiable in a domain $\mathfrak{G}$ of the (x, y)-plane is called *harmonic* in $\mathfrak{G}$ if it satisfies throughout $\mathfrak{G}$ the potential equation

$$u_{xx} + u_{yy} = 0.$$

As we are about to show, there is a simple connection between harmonic and analytic functions. Let $z = x + iy$ and let $f(z) = u + iv$ be a regular analytic function of z in $\mathfrak{G}$. Then the real and imaginary parts $u(z)$ and $v(z)$ of $f(z)$ are infinitely differentiable in $\mathfrak{G}$ and satisfy the Cauchy-Riemann equations

$$u_x = v_y, \qquad u_y = -v_x,$$

so that

$$u_{xx} + u_{yy} = v_{yx} - v_{xy} = 0, \qquad v_{xx} + v_{yy} = -u_{yx} + u_{xy} = 0.$$

This means that the real and imaginary parts of a regular analytic function in $\mathfrak{G}$ are harmonic in $\mathfrak{G}$. The problem now is to find an analytic function whose real part is a preassigned harmonic function.

Theorem 2: Let u be a harmonic function in $\mathfrak{G}$. Then there exists a regular but not necessarily single-valued function $f(z)$ in $\mathfrak{G}$ whose real part is u and whose imaginary part is determined to within an additive constant.

Proof: We must find a real-valued function $v(z)$ in $\mathfrak{G}$ whose two partial derivatives have in $\mathfrak{G}$ the prescribed values

$$v_x = -u_y = A, \qquad v_y = u_x = B.$$

A necessary and sufficient condition for the integrability of these two differential equations is that the relation

$$A_y = B_x$$

is satisfied locally. But this condition holds in view of the assumption that u satisfies the potential equation. The solution, which is determined to within an additive constant, is given by the line integral

$$I(z) = \int_{z_0}^{z} (A \, dx + B \, dy),$$

where the path of integration is a curve in $\mathfrak{G}$ from a fixed initial point z_0 to a variable terminal point z. If the terminal point is kept fixed and we admit only homotopic paths, then, surely the integral is path-independent; in particular, the integral is path-independent for a simply connected domain $\mathfrak{G}$. If c is any real constant of integration then $v(z) = I(z) + c$ is said to be conjugate harmonic to u. It must be stressed that if the region is multiply connected, then the function may exhibit additive multiple-valuedness corresponding to the periods of the integral. In that case, the function $f(z) = u + iv$ while regular under analytic continuation in $\mathfrak{G}$, need not be single-valued; but then two different branches differ by a pure imaginary constant only. This completes the proof. Incidentally, this proof can be given the following somewhat different form. The complex function

$$g(z) = u_x - iu_y$$

has the real part u_x and the imaginary part $-u_y$. In view of the potential function for u, these functions satisfy the Cauchy-Riemann equations

$$(u_x)_x = (-u_y)_y, \qquad (u_x)_y = (-u_y)_x.$$

Thus $g(z)$ is a single-valued and regular function of z in $\mathfrak{G}$. Integration of $g(z)$ yields an analytic function $f(z)$ determined to within an arbitrary complex constant of integration, for which

$$f_x = f_z = g, \qquad f_y = if_z = ig,$$

so that the real part of f has the partial derivatives u_x and y_y. It follows that the real part of f differs from u by at most an additive constant which we may choose to be 0. To shed light on the possible multiple-valuedness of the imaginary part of f we consider the two simple examples $u = x$ and $u = \frac{1}{2} \log (x^2 + y^2)$. In the first case we take $\mathfrak{G}$ to be the whole plane and we obtain $g(z) = 1, f(z) = z + ic$. In the second case we take $\mathfrak{G}$ to be the plane punctured at $z = 0$, and we obtain

$$g(z) = \frac{x}{x^2 + y^2} - \frac{iy}{x^2 + y^2} = z^{-1}, \qquad f(z) = \log z + ic.$$

We see that $f(z)$ is single-valued in $\mathfrak{G}$ only in the first of these two examples.

According to Theorem 2 every harmonic function is the real part of a regular analytic function. This implies that it is not just twice differentiable but infinitely differentiable.

Let $\mathfrak{H}$ be a schlicht domain in the (x, y)-plane containing $\mathfrak{G}$ and let u be a harmonic function in $\mathfrak{G}$. If there exists a harmonic function in $\mathfrak{H}$ which coincides with u on $\mathfrak{G}$, then this function is called a *harmonic continuation* of u. Theorem 2 implies that if such a continuation is at all possible, then it is unique. Indeed, the possibility of harmonic continuation implies that there exists a regular analytic function in $\mathfrak{H}$ whose real part yields the harmonic continuation. By the principle of analytic continuation this function is determined by its values on $\mathfrak{G}$ and, therefore also, up to a pure imaginary constant, by u. This result reduces the problem of harmonic continuation of a harmonic function to the problem of analytic continuation of $f(z) = u + iv$.

Theorem 3: Suppose that the transformation

$$x = \varphi(p, q), \qquad y = \psi(p, q)$$

maps a domain $\mathfrak{F}$ in the (p, q)-plane conformally onto a domain $\mathfrak{G}$ in the (x, y)-plane. Then every function of x and y harmonic in $\mathfrak{G}$ is also harmonic in $\mathfrak{F}$ when viewed as a function in the variables p and q.

Proof: If the conformal mapping preserves orientation, then $x + iy$, considered in $\mathfrak{F}$, is a regular analytic function of the complex variable $p + iq$. With the obvious restriction on the domains of definition, an analytic function of an analytic function is an analytic function. Now our assertion follows from Theorem 2. The case when the conformal mapping reverses orientation is reduced to the case just treated by transition from z to $\bar{z}$; here the essential point is that this substitution does not alter the Laplacian $u_{xx} + u_{yy}$.

Let C be a simple closed rectifiable curve in the complex plane, and let $f(z)$ be a function regular in the interior and on the boundary of C. In view of the Cauchy integral formula we have

$$f(z) = \frac{1}{2\pi i} \int_C \frac{f(\zeta)}{\zeta - z} \, d\zeta,$$

where we integrate in the positive sense if the point z is in the interior of C. We wish to formulate an analogue of this result for harmonic functions. In particular, we suppose C a circle of radius ρ. The center of C may be taken to be at the origin; this may require the use of an appropriate translation. In the sequel we shall have occasion to use the abbreviation

$$P(z, \zeta) = \frac{\zeta\bar{\zeta} - z\bar{z}}{(\zeta - z)(\bar{\zeta} - \bar{z})}.$$

Theorem 4: Let $u(z)$ be harmonic in a domain containing the closed disk $|z| \leq \rho$. With C positively oriented we have

$$u(z) = \frac{1}{2\pi} \int_C P(z, \zeta) u(\zeta) \frac{d\zeta}{i\zeta} \qquad (|z| < \rho).$$

Proof: If ζ is on the boundary of C and z is in its interior, then

$$\zeta\bar{\zeta} - z\bar{z} > 0, \qquad (\zeta - z)(\bar{\zeta} - \bar{z}) > 0,$$

so that $P(z, \zeta)$ is real and positive. We now choose a radius $R > \rho$ such that $u(z)$ is harmonic in the disk $|z| < R$ and take $f(z)$ to be the analytic function, determined up to an additive imaginary constant, whose real part is $u(z)$. In particular, this function is regular in the closed disk $|z| \leq \rho$ and, by the Cauchy integral formula, we have

$$(2) \qquad f(z) = \frac{1}{2\pi} \int_C \frac{\zeta}{\zeta - z} f(\zeta) \frac{d\zeta}{i\zeta} \qquad (|z| < \rho).$$

Further, we consider the point

$$z^* = \frac{\rho^2}{\bar{z}}$$

obtained by reflecting the point z in the circle C. Since z^* lies outside the circle, the function $(\zeta - z^*)^{-1} f(\zeta)$ of ζ is regular on the closed disk $|\zeta| \leq \rho$ and so, by Cauchy's theorem,

$$(3) \qquad 0 = \frac{1}{2\pi} \int_C \frac{\zeta}{\zeta - z^*} f(\zeta) \frac{d\zeta}{i\zeta} \qquad (|z| < \rho).$$

From $\zeta\bar{\zeta} = z^*\bar{z} = \rho^2$ we have

$$\frac{\zeta}{\zeta - z} - \frac{\zeta}{\zeta - z^*} = \frac{\zeta(z - z^*)}{(\zeta - z)(\zeta - z^*)} = \frac{\zeta(z\bar{z} - \zeta\bar{\zeta})}{(\zeta - z)(\zeta\bar{z} - \zeta\bar{\zeta})} = P(z, \zeta).$$

From (2) and (3) we obtain by subtraction

$$(4) \qquad f(z) = \frac{1}{2\pi} \int_C P(z, \zeta) f(\zeta) \frac{d\zeta}{i\zeta}.$$

If we introduce polar coordinates by means of the substitution $\zeta = \rho e^{i\varphi}$, then

$$\frac{d\zeta}{i\zeta} = d\varphi$$

is real. The proof of our theorem follows if we form the real part in (4).

The equation of Theorem 4 is called the *Poisson integral formula* and the function $P(z, \zeta)$, the *Poisson kernel*, in honor of the French mathematician Poisson who was a contemporary of Cauchy. From this formula we now deduce the maximum principle for harmonic functions in a manner similar to that used in deducing the maximum principle for analytic functions from the Cauchy integral formula. This is achieved in three steps.

Theorem 5: Let $u(z)$ be harmonic in a domain containing the closed disk $|z| \leq \rho$. If m is the maximum of $u(\zeta)$ on the circle $|\zeta| = \rho$, then $u(z) \leq m$ for z in the interior of the disk; better still, if $u(z)$ is not constant in the domain then $u(z) < m$.

Proof: The Poisson integral formula

$$u(z) = \frac{1}{2\pi} \int_0^{2\pi} P(z, \zeta) u(\zeta) \, d\varphi \qquad (\zeta = \rho e^{i\varphi}, |z| < \rho)$$

reduces in the special case $u = m$ to

$$m = \frac{1}{2\pi} \int_0^{2\pi} P(z, \zeta) m \, d\varphi.$$

From this we obtain by subtraction

$$(5) \qquad m - u(z) = \frac{1}{2\pi} \int_0^{2\pi} P(z, \zeta)(m - u(\zeta)) \, d\varphi \qquad (|z| < \rho).$$

In the integrand we have throughout $P(z, \zeta) > 0$ and, by assumption, $m - u(\zeta) \geq 0$, so that, by the mean value theorem,

$$m - u(z) \geq 0 \qquad (|z| < \rho).$$

If for some value of z the inequality could be replaced by an equality, then the integral in (5) would have the value 0. In view of the continuity of the positive function $P(z, \zeta)$ and the nonnegative function $m - u(\zeta)$ this occurs if and only if $m - u(\zeta)$ vanishes at all points of the circle, in which case $m - u(z) = 0$ in its interior. Using harmonic continuation we conclude that in the exceptional case $u(z)$ has the constant value m. Our result can be extended to a disk with arbitrary center by applying a suitable translation.

Theorem 6: A nonconstant function harmonic in a domain has no local extremum in the domain.

Proof: Let $u(z)$ be harmonic in a domain $\mathfrak{G}$ and let ζ be a point at which $u(z)$ has a local maximum M. We take a disk about ζ which is so small that its closure is in $\mathfrak{G}$ and $u(z) \leq M$ at all the points of the closure. The maximum m of $u(z)$ on the boundary of the disk is also $\leq M$. On the other hand, $u(\zeta) = M$ at the center ζ. In view of Theorem 5 we arrive at the contradiction that $u(z)$ is constant. Since the harmonicity of u implies the harmonicity of $-u$, our conclusion holds in the case of a local minimum. This completes the proof.

Theorem 7: Consider a function which is harmonic in a bounded plane domain and has boundary values throughout. Such a function takes on its absolute extremal values on the boundary of the domain.

Proof: Let $\tilde{\mathfrak{G}}$ be the result of adjoining to $\mathfrak{G}$ its boundary points. In view of Theorem 1, the function $u(z)$ extended by means of its boundary values and harmonic in $\mathfrak{G}$ is continuous on $\tilde{\mathfrak{G}}$. Since $\tilde{\mathfrak{G}}$ is compact there is a point ζ in $\tilde{\mathfrak{G}}$ at which u takes on an absolute maximum. A similar statement is true of the minimum. If u is not constant then Theorem 6 implies that ζ is a boundary point of $\mathfrak{G}$. If u is constant then ζ can be chosen to be a boundary point. This completes the proof.

With Theorem 7 established, let m denote the maximum of the boundary values of the function $u(z)$. Then $u(z) \leq m$ in the interior of the domain and, better still, $u(z) < m$ for $u(z)$ not a constant. This is the maximum principle for harmonic functions.

Theorem 8: Consider a function which is harmonic in a bounded plane domain and has boundary values throughout. Then there is no other function harmonic in the domain and having the same boundary values.

Proof: Let u_1 and u_2 be harmonic functions in a bounded domain $\mathfrak{G}$ with the same boundary values. Then the function $u = u_1 - u_2$ is harmonic in $\mathfrak{G}$ and vanishes on the boundary of $\mathfrak{G}$. By Theorem 7, $0 \leq u \leq 0$ in all of $\mathfrak{G}$, so that $u = 0$ and $u_1 = u_2$ as asserted.

Theorem 8 states that a function harmonic in a domain $\mathfrak{G}$ is uniquely determined by its boundary values provided the latter exist throughout, and the domain is bounded. The boundary value problem of potential theory can be stated as follows: given a continuous function $g(\zeta)$ on the boundary of $\mathfrak{G}$, find a function which is harmonic in $\mathfrak{G}$ and takes on the prescribed boundary values $g(\zeta)$. We first solve this boundary value problem for a disk which we assume, with no loss of generality, is centered at the origin.

Theorem 9: Let C be a positively oriented circle with radius ρ and center at the origin and let $g(\zeta)$ be a real-valued function continuous on C. Then the integral

$$u(z) = \frac{1}{2\pi} \int_C P(z, \zeta) g(\zeta) \frac{d\zeta}{i\zeta} \qquad (|z| < \rho)$$

represents a function which is harmonic in the interior of C and has the boundary values $g(\zeta)$.

Proof: For the Poisson kernel we have the identity

$$P(z, \zeta) = \frac{\zeta\bar{\zeta} - z\bar{z}}{(\zeta - z)(\bar{\zeta} - \bar{z})} = \left(\frac{\zeta}{\zeta - z} - \frac{1}{2}\right) + \left(\frac{\bar{z}}{\bar{\zeta} - \bar{z}} - \frac{1}{2}\right).$$

If we put

$$f(z) = \frac{1}{\pi} \int_C \left(\frac{\zeta}{\zeta - z} - \frac{1}{2} \right) g(\zeta) \frac{d\zeta}{i\zeta} \qquad (|z| < \rho),$$

then we have

(6)
$$u(z) = \frac{f(z) + \overline{f(z)}}{2}.$$

On the other hand, consider the geometric series

$$\frac{\zeta}{\zeta - z} = \sum_{n=0}^{\infty} z^n \zeta^{-n}$$

for a fixed $|z| < \rho$ and $|\zeta| = \rho$. Since $|z\zeta^{-1}| = |z|\rho^{-1}$, this series converges on C uniformly in ζ, so that we may interchange the order of summation and integration. In this way we obtain the convergent power series

(7)
$$f(z) = \tfrac{1}{2}a_0 + \sum_{n=1}^{\infty} a_n z^n \qquad (|z| < \rho)$$

with

(8)
$$a_n = \frac{1}{\pi} \int_C \zeta^{-n} g(\zeta) \frac{d\zeta}{i\zeta} \qquad (n = 0, 1, \ldots).$$

The function $u(z)$ is harmonic in the interior of C because it is the real part of the function $f(z)$, which is regular analytic there.

Let ζ_0 be a point of C and let z vary over a sequence of values converging to ζ_0 with $|z| < \rho$. We must show that $u(z)$ converges to the boundary value $g(\zeta_0)$. In view of the continuity of $g(\zeta)$ we can find for every $\varepsilon > 0$ a number $\delta(\varepsilon) = \delta > 0$ such that

$$|g(\zeta) - g(\zeta_0)| < \varepsilon \qquad (|\zeta - \zeta_0| < \delta).$$

Let C_1 be the arc of C with $|\zeta - \zeta_0| < \delta$, and let C_2 be the remaining arc of C. We shall obtain estimates for the two integrals

$$I_k = \frac{1}{2\pi} \int_{C_k} P(z, \zeta)(g(\zeta) - g(\zeta_0)) \frac{d\zeta}{i\zeta} \qquad (k = 1, 2).$$

Application of Theorem 4 to the particular function $u(z) = 1$ yields

$$\frac{1}{2\pi} \int_C P(z, \zeta) \frac{d\zeta}{i\zeta} = 1,$$

and therefore

$$|I_1| < \frac{1}{2\pi} \int_{C_1} P(z, \zeta)\varepsilon \frac{d\zeta}{i\zeta} \le \frac{\varepsilon}{2\pi} \int_C P(z, \zeta) \frac{d\zeta}{i\zeta} = \varepsilon.$$

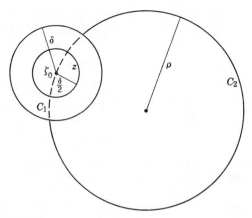

Figure 55

Further, the function $P(z, \zeta)$ is discontinuous only at $z = \zeta$. If ζ varies on C_2 and z is subject to the restriction $|z - \zeta_0| < \delta/2$, then $P(z, \zeta)$ is uniformly continuous there and, for $z \to \zeta_0$, $P(z, \zeta)$ converges uniformly in ζ to the value $P(\zeta_0, \zeta) = 0$ (Figure 55). In view of the boundedness of $g(\zeta)$ we have, for z sufficiently close to ζ_0, $|I_2| < \varepsilon$. This shows that as $z \to \zeta_0$, $I_1 + I_2 \to 0$. It follows that

$$u(z) = \frac{1}{2\pi} \int_C P(z, \zeta) g(\zeta) \frac{d\zeta}{i\zeta} \to \frac{1}{2\pi} \int_C P(z, \zeta) g(\zeta_0) \frac{d\zeta}{i\zeta} = g(\zeta_0)$$

as $z \to \zeta_0$, which completes the proof of Theorem 9. The integral for $u(z)$ is known as the *Poisson integral* and it yields the solution of the boundary value problem of potential theory for a disk.

Now we come to Harnack's theorem for harmonic functions. This theorem is the analogue of the well-known Weierstrass theorem for sequences of analytic functions.

Theorem 10: Let $u_n(z)$ $(n = 1, 2, \ldots)$ be a sequence of harmonic functions defined in a bounded plane domain $\mathfrak{G}$ all of which have boundary values throughout. Suppose that the sequence $u_n(\zeta)$ converges for all boundary points ζ and that this convergence is uniform on the whole boundary. Then $u_n(z)$ converges uniformly in $\mathfrak{G}$ to a harmonic function $u(z)$ which also has boundary values throughout; specifically

$$u(\zeta) = \lim_{n \to \infty} u_n(\zeta).$$

If $\mathfrak{F}$ is a subdomain of $\mathfrak{G}$ whose boundary lies in $\mathfrak{G}$, then, for $n \to \infty$, every partial derivative of $u_n(z)$ of arbitrary order converges uniformly on $\mathfrak{F}$ to the corresponding partial derivative of $u(z)$.

Proof: The assumed uniform convergence of $u_n(\zeta)$ means that we can find for every positive ε a number $N(\varepsilon) = N$ independent of ζ such that the inequalities

$$-\varepsilon < u_n(\zeta) - u_m(\zeta) < \varepsilon \qquad (n, m > N)$$

hold on the whole boundary of $\mathfrak{G}$. Since the difference $u_n(z) - u_m(z)$ is likewise harmonic in $\mathfrak{G}$, we surely have by Theorem 7

$$-\varepsilon < u_n(z) - u_m(z) < \varepsilon \qquad (n, m > N)$$

for all z in $\mathfrak{G}$. This proves the uniform convergence of the sequence $u_n(z)$ to a limit function $u(z)$. In fact, this assertion holds on the closed domain $\tilde{\mathfrak{G}}$. Consequently, $u(z)$ is continuous there and has, throughout, the required boundary values.

Now we show that $u(z)$ is harmonic in $\mathfrak{G}$. Let z_0 be a point of $\mathfrak{G}$ which we suppose shifted to 0 by a translation. About this point we form a circle C with radius ρ such that the closed disk $|z| \leq \rho$ lies entirely in $\mathfrak{G}$. In view of the Poisson integral formula of Theorem 4 we have

$$u_n(z) = \frac{1}{2\pi} \int_C P(z, \zeta) u_n(\zeta) \frac{d\zeta}{i\zeta} \qquad (n = 1, 2, \ldots),$$

for $|z| < \rho$, where ζ is now the variable of integration. The sequence $u_n(\zeta)$ $(n = 1, 2, \ldots)$ converges on C uniformly to $u(\zeta)$ and therefore

$$u(z) = \frac{1}{2\pi} \int_C P(z, \zeta) u(\zeta) \frac{d\zeta}{i\zeta} \qquad (|z| < \rho).$$

In view of Theorem 9, $u(z)$ is seen to be harmonic in the interior of C. Since z_0 was arbitrary, $u(z)$ is harmonic in all of $\mathfrak{G}$ as asserted.

For $z \neq \zeta$ the Poisson kernel $P(\zeta, z)$ has partial derivatives of all orders with respect to x and y. It follows that in the last two formulas it is possible to obtain all the derivatives of $u_n(z)$ and $u(z)$ with respect to x and y by differentiating under the integral sign. Since $u_n(\zeta)$ converges uniformly on C to $u(\zeta)$, the resulting formulas imply that every partial derivative of $u_n(z)$ converges uniformly to the corresponding derivative of $u(z)$ in the interior of every circle concentric with and smaller than C. The fact that $\mathfrak{F}$ can be covered by finitely many such disks justifies the remainder of our theorem.

6. The Dirichlet integral

Let $\mathfrak{G}$ be a bounded domain in the (x, y)-plane having Jordan content. Further, let v be a continuous real valued function on $\mathfrak{G}$ with continuous derivatives v_x and v_y. If these derivatives are bounded, then the integral

$$\iint_{\mathfrak{G}} (v_x^2 + v_y^2) \, dx \, dy = D[v; \mathfrak{G}]$$

exists and is called the *Dirichlet integral* of v over $\mathfrak{G}$. If not both of the derivatives v_x and v_y are bounded, then, in the case of convergence the integral is defined in the usual way as an improper integral. Finally, if $\tilde{\mathfrak{G}}$ is the result of adding to $\mathfrak{G}$ all of its boundary points, then we put

$$D[v; \tilde{\mathfrak{G}}] = D[v; \mathfrak{G}].$$

Throughout the remainder of this section we shall consider only closed Jordan domains and will refer to them briefly as regions.

Let $\mathfrak{B}_1, \ldots,$ be $\mathfrak{B}_n$ non-overlapping regions, and let $\mathfrak{B}$ stand for the region which is their union. If v is a continuous function defined on $\mathfrak{B}$ with continuous partial derivatives of the first order in the interior of each subregion $\mathfrak{B}_k$ $(k = 1, \ldots, n)$, then v is said to be piecewise differentiable. The Dirichlet integral of v over $\mathfrak{B}$ is defined by

$$D[v; \mathfrak{B}] = \sum_{k=1}^{n} D[v; \mathfrak{B}_k],$$

where the individual summands are supposed finite. It is easy to show that the value of $D[v; \mathfrak{B}]$ is independent of the decomposition into admissible subregions.

Theorem 1: Let a domain $\mathfrak{F}$ in the (p, q)-plane be mapped conformally onto a domain $\mathfrak{G}$ in the (x, y)-plane and suppose that this mapping carries a region $\mathfrak{A}$ in $\mathfrak{F}$ to the region $\mathfrak{B}$. If v is piecewise differentiable function of x, y in $\mathfrak{B}$, then v is also piecewise differentiable when viewed as a function of p, q in $\mathfrak{A}$ and we have

$$D[v; \mathfrak{B}] = D[v; \mathfrak{A}].$$

Proof: In view of the conformality of the mapping, the derivatives x_p, x_q, y_p, y_q are continuous on $\mathfrak{F}$ and

$$x_p = y_q, \qquad x_q = -y_p,$$

provided that the orientation is preserved. Therefore

$$v_p = v_x x_p + v_y y_p = v_x x_p - v_y x_q, \qquad v_q = v_x x_q + v_y y_q = v_x x_q + v_y x_p,$$

$$v_p^2 + v_q^2 = (v_x^2 + v_y^2)(x_p^2 + x_q^2), \qquad \frac{d(x, y)}{d(p, q)} = x_p y_q - x_q y_p = x_p^2 + x_q^2,$$

$$(v_p^2 + v_q^2)\, dp\, dq = (v_x^2 + v_y^2)\, dx\, dy.$$

Consideration of the inverse mapping implies that every subregion $\mathfrak{B}_k$ again corresponds to a Jordan subregion $\mathfrak{A}_k$ on which the derivatives v_p an v_q are continuous. It follows that

$$D[v; \mathfrak{A}_k] = D[v, \mathfrak{B}_k].$$

This implies our theorem for the case when the mapping preserves orientation In the case of an orientation reversing mapping we need only introduce an additional reflection in the x-axis.

Theorem 2: Let u and v be two real valued functions which are continuously differentiable in the interior of a region $\mathfrak{B}$, and suppose that the two Dirichlet integrals $D[u, \mathfrak{B}]$ and $D[v, \mathfrak{B}]$ exist. Then the integrals $D[\lambda u + \mu v; \mathfrak{B}]$, where λ, μ are real constants, and the integral

$$\iint_{\mathfrak{B}} (u_x v_x + u_y v_y) \, dx \, dy = D[u, v; \mathfrak{B}]$$

all exist, and we have the relation

$$D[\lambda u + \mu v; \mathfrak{B}] = \lambda^2 D[u; \mathfrak{B}] + 2\lambda\mu D[u, v; \mathfrak{B}] + \mu^2 D[v; \mathfrak{B}].$$

Proof: We have

$$|u_x v_x + u_y v_y| \leq |u_x v_x| + |u_y v_y| \leq \tfrac{1}{2}(u_x^2 + u_y^2 + v_x^2 + v_y^2).$$

In view of the existence of $D[u; \mathfrak{B}]$ and $D[v; \mathfrak{B}]$, this implies the absolute convergence of the integral $D[u, v; \mathfrak{B}]$. To establish the remaining part of the theorem we put $\lambda u + \mu v = w$ and note that

$$w_x^2 + w_y^2 = \lambda^2(u_x^2 + u_y^2) + 2\lambda\mu(u_x v_x + u_y v_y) + \mu^2(v_x^2 + v_y^2).$$

Now, more generally, let u and v be piecewise differentiable in $\mathfrak{B}$ and let there exist a decomposition of $\mathfrak{B}$ into subregions $\mathfrak{B}_k$ $(k = 1, \ldots, n)$ such that u and v have continuous derivatives of the first order in the interior of each $\mathfrak{B}_k$. Then, clearly, Theorem 2 holds under this weaker assumption provided that we define

$$D[u, v; \mathfrak{B}] = \sum_{k=1}^{n} D[u, v; \mathfrak{B}_k].$$

Using Theorem 1 we can readily prove the invariance of $D[u, v; \mathfrak{B}]$ under conformal mapping.

After these simple consequences of the definition of the Dirichlet integral we come to deeper theorems which bear on the connection between the Dirichlet integral and harmonic functions. The functions considered in the remainder of this section are, as before, real valued. It is necessary, however, to restrict the notion of a region still further. Specifically, we assume from now on that the boundaries of $\mathfrak{B}$ and of the subregions $\mathfrak{B}_k$ are piecewise smooth. In other words, each boundary consists of finitely many simple curves each of which admits of a continuously differentiable parametric representation $x = f(s)$, $y = g(s)$, with s the arc length. We denote the positively oriented boundary of $\mathfrak{B}$ by C. This boundary may consist of a number of simple closed curves, since $\mathfrak{B}$ need not be simply connected. We use the customary abbreviation

$$\Delta u = u_{xx} + u_{yy}.$$

Theorem 3: Let u and v be continuous on $\mathfrak{B}$. We assume that the derivatives u_x, u_y, u_{xx}, u_{yy} exist and are continuous in the interior of $\mathfrak{B}$ and that they have boundary values throughout. Of v we assume only that it is piecewise differentiable and that the Dirichlet integral $D[v; \mathfrak{B}]$ exists. Then we have the relation

(1) $$D[u, v; \mathfrak{B}] + \iint_{\mathfrak{B}} (\Delta u)v \, dx \, dy = \int_C v(u_x \, dy - u_y \, dx).$$

Proof: The assertion is called *Green's identity* and is obtained from the well-known formula of the integral calculus which is variously associated with the names of Green, Gauss, and Riemann. It asserts that

(2) $$\iint_{\mathfrak{B}} (q_x - p_y) \, dx \, dy = \int_C (p \, dx + q \, dy)$$

when p, q, p_y, q_x are continuous in the interior of $\mathfrak{B}$ and have boundary values throughout. If we put $p = -u_y v$, $q = u_x v$ then we have

$$q_x = u_x v_x - u_{xx} v, \quad -p_y = u_y v_y + u_{yy} v, \quad q_x - p_y = (u_x v_x + u_y v_y) + (\Delta u)v,$$

and this implies the assertion for the special case when the derivatives v_x, v_y are continuous in the interior of $\mathfrak{B}$ and have boundary values throughout.

In the sequel we shall require the theorem under the assumption formulated above, namely, that v is only piecewise differentiable and the Dirichlet integral exists. We consider first the case when v_x and v_y are continuous in the interior of $\mathfrak{B}$, but we do not assume the existence of boundary values on C. In particular, v_x and v_y need not be bounded in $\mathfrak{B}$. Under these circumstances we may not use (2) directly, but must first approximate the boundary C in the following manner. We decompose C into finitely many smooth subarcs $C_1, \ldots, C_m$ such that on each of them the direction of the tangent varies by less than $\pi/2$. If L_k $(k = 1, \ldots, m)$ denotes the inner normal at the initial point a_k of the boundary arc C_k, then, according to the mean value theorem of the differential calculus, no parallel to L_k meets the curve C_k in more than one point. We draw about the points a_k circles D_k of radius ε, which is so small that these circles are disjoint and D_k intersects the boundary C exactly once at a point of C_k and once at a point of the preceding arc. Next we translate C_k in the direction L_k by a distance $\delta < \varepsilon$, and denote the resulting arc by C_k^*. If $\delta < \delta_0(\varepsilon)$, then each of the two circles about the end points of C_k also meets the arc C_k^* in exactly one point and determines on C_k^* a middle portion C_k' which lies entirely in $\mathfrak{B}$ and meets neither the remaining circles

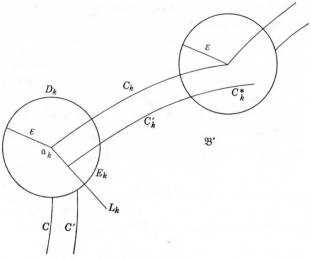

Figure 56

nor the other arcs C'_l ($l \neq k$). If we join the C'_k in pairs by means of arcs E_k of the D_k, then we obtain the boundary C' of a region $\mathfrak{B}'$ in the interior of $\mathfrak{B}$ (Figure 56) to which we may apply the results of the preceding paragraph. We have, accordingly,

$$D[u, v; \mathfrak{B}'] + \iint_{\mathfrak{B}'} (\Delta u)v \, dx \, dy = \int_{C'} v(u_x \, dy - u_y \, dx).$$

Now we let ε go to 0. In view of the existence of $D[u, \mathfrak{B}]$ and $D[v; \mathfrak{B}]$, the integral $D[u, v; \mathfrak{B}']$ converges to $D[u, v; \mathfrak{B}]$. Since, in addition, the functions $(\Delta u)v$, vu_x, $-vu_y$ are continuous on $\mathfrak{B}$, and C_k^* is obtained from C_k by a translation, it follows that in passing to the limit the double integral and the line integral converge to the appropriate integrals in (1). This settles the case now under consideration.

In the general case we must divide $\mathfrak{B}$ into subregions $\mathfrak{B}_1, \ldots, \mathfrak{B}_n$ for which the assumption concerning v_x and v_y just used holds, and apply the already established result. Then addition yields the general result provided that we bear in mind that in the case of the line integrals the integration over the parts of the boundaries of the $\mathfrak{B}_k$ in the interior of $\mathfrak{B}$ is always carried out twice, in opposite directions, and therefore does not affect the final result.

The following is an important consequence of Theorem 3. Let $g(\zeta)$ be a given continuous functions on the boundary C of $\mathfrak{B}$. We consider all functions which are piecewise differentiable in $\mathfrak{B}$, take on the prescribed boundary values $g(\zeta)$, and have a Dirichlet integral. The following theorem shows that the harmonic functions occupy a special position in this class of functions.

Theorem 4: Let u be a function harmonic in the interior of $\mathfrak{B}$ such that u and its derivatives u_x, u_y, u_{xx}, u_{yy} have boundary values throughout. If w is a function which is continuous and piecewise differentiable on $\mathfrak{B}$ with a convergent Dirichlet integral, and if w coincides with u on the boundary, then

$$D[u; \mathfrak{B}] \leq D[w; \mathfrak{B}].$$

Here we have equality only if $u = w$ throughout $\mathfrak{B}$.

Proof: We put $w - u = v$ and apply Theorem 3, as we may, to the two functions u and v. Since v vanishes on the boundary C and $\Delta u = 0$ in the interior of $\mathfrak{B}$, we have

$$D[u, v; \mathfrak{B}] = 0.$$

In view of Theorem 2 we have

$$D[w; \mathfrak{B}] = D[u + v; \mathfrak{B}] = D[u; \mathfrak{B}] + D[v, \mathfrak{B}] \geq D[u; \mathfrak{B}].$$

Here we may replace the inequality with an equality if and only if

$$0 = D[v; \mathfrak{B}] = \sum_{k=1}^{n} \iint\limits_{\mathfrak{G}_k} (v_x^2 + v_y^2) \, dx \, dy,$$

where $\mathfrak{G}_k$ denotes the interior of the appropriately defined subregion $\mathfrak{B}_k$. Since v_x and v_y are continuous, v must be constant in $\mathfrak{G}_k$. On the other hand, v is continuous in $\mathfrak{B}$ and 0 on the boundary. It follows that $v = 0$ and $u = w$ on $\mathfrak{B}$. This completes the proof.

We wish to formulate Theorem 4 differently. We consider all functions w which are continuous and piecewise differentiable in $\mathfrak{B}$, have a convergent Dirichlet integral and take on the prescribed boundary values $g(\zeta)$, and pose the problem of minimizing the Dirichlet integral $D[w; \mathfrak{B}]$ by a suitable choice of w. If there exists a function u harmonic in the interior of $\mathfrak{B}$ which takes on the prescribed boundary values $g(\zeta)$ and which, together with its derivatives u_x, u_y, u_{xx}, u_{yy}, has boundary values throughout, then this harmonic function has the required minimizing property and is uniquely determined by this property in the class of admissible functions w. So far we solved the boundary value problem of potential theory only for the special case of a disk (cf. Theorem 9 in Section 5), and even in that special case we have not proved the existence of boundary values of the derivatives. Examples show that, in general, the derivatives of a harmonic function need not have boundary values.

The solvability of the boundary value problem for an arbitrary region $\mathfrak{B}$ is made plausible by an argument due to Klein. We think of the surface $\mathfrak{B}$ as a heat conducting layer which has the same conducting ability everywhere, and denote by $g(\zeta)$ the time independent temperature at the boundary point

ζ. Due to the different temperatures at the boundary we obtain in $\mathfrak{B}$ a stationary heat flow. If $u = u(z)$ denotes the temperature at the point z of $\mathfrak{B}$, then it follows from certain physical assumptions that u is harmonic in the interior of $\mathfrak{B}$ and has the boundary values $g(\zeta)$. If v is the conjugate harmonic function, then, for constant values of c, the curves $v = c$ are just the heat flow lines and their orthogonal trajectories are the isothermal lines $u = c$. This train of thought suggests that it is reasonable to expect the boundary value problem of potential theory to be solvable for an arbitrary region $\mathfrak{B}$. Riemann tried to obtain a rigorous solution of this problem. His approach aimed at the determination for the prescribed boundary values of a function which minimizes the Dirichlet integral. Before discussing the difficulties of the relevant existence proof we assume that the extremum problem has a solution u whose derivatives u_x, u_y, u_{xx}, u_{yy} are continuous in the interior of $\mathfrak{B}$ and have boundary values throughout and prove that, under these conditions, u is the solution of the boundary value problem.

Theorem 5: Let u be a function corresponding to the prescribed boundary values which is continuous and piecewise differentiable on $\mathfrak{B}$, and minimizes the Dirichlet integral. If the derivatives u_x, u_y, u_{xx}, u_{yy} exist in the interior of $\mathfrak{B}$, are continuous there and have boundary values throughout, then u is harmonic in the interior of $\mathfrak{B}$.

Proof: The proof follows the pattern common in the calculus of variations. Let v be a function which is continuous and piecewise differentiable in $\mathfrak{B}$, has a convergent Dirichlet integral and vanishes on the boundary. Let $w = u + \varepsilon v$, where ε is an arbitrary real number. Since this function has the same boundary values as u, it follows, by assumption, that

$$(3) \qquad\qquad D[w; \mathfrak{B}] \geq D[u; \mathfrak{B}].$$

On the other hand, Theorem 2 yields

$$D[w; \mathfrak{B}] - D[u; \mathfrak{B}] = \varepsilon(2D[u, v; \mathfrak{B}] + \varepsilon D[v; \mathfrak{B}]).$$

If in this last relation $D[u, v; \mathfrak{B}] \neq 0$, then ε could be chosen so that, contrary to (3), the expression on the right is negative. It follows that

$$D[u, v; \mathfrak{B}] = 0.$$

Since, in addition, the function v vanishes on the boundary, we have

$$(4) \qquad\qquad \iint_{\mathfrak{B}} (\Delta u) v \, dx \, dy = 0$$

as a consequence of Theorem 3. If $\Delta u \neq 0$ at some interior point z_0 of $\mathfrak{B}$, then, in view of the continuity of u_{xx} and u_{yy}, the function Δu would have a fixed sign in some disk in $\mathfrak{B}$ with sufficiently small radius ρ and center z_0. We put

$$v(z) = \rho^2 - |z - z_0|^2$$

in this disk and $v(z) = 0$ in the remainder of $\mathfrak{B}$. In view of the mean value theorem of the integral calculus, the left-hand side of (4) would not be 0 but would have the sign of Δu at z_0. Since v satisfies the relevant assumptions this would lead to a contradiction. It follows that $\Delta u = 0$ in the interior of $\mathfrak{B}$. This completes the proof.

An essential part of Theorem 5 is the assumed existence of a function u which, under given boundary values, minimizes the Dirichlet integral. Riemann assumed the existence of such an extremum function without proof. Riemann took over this argument from Dirichlet, and this has earned it the name of the *Dirichlet principle*. Before that Gauss used the same argument. Weierstrass was critical of it and based his objections on the fact that even very simple variational problems need not have a solution. Many decades later Hilbert succeeded in justifying the Dirichlet principle. In this connection it should be pointed out that in the proof of Theorem 3 we made use of Green's identity and this necessitates the assumptions on u_x, u_y, u_{xx}, u_{yy}; these assumptions make an existence proof particularly difficult. In Section 8 we shall give a detailed treatment of a closely related extremum problem. Here we consider this issue only for a disk.

Theorem 6: Let $\mathfrak{B}$ be a disk of radius ρ about the origin and let $g(\zeta)$ be a given continuous function on the boundary C of the disk. Among all functions which are continuous and piecewise differentiable on $\mathfrak{B}$, take on the boundary values $g(\zeta)$, and have a convergent Dirichlet integral (we assume the existence of such a function) only the harmonic function given by the Poisson integral

$$u(z) = \frac{1}{2\pi} \int_C P(z, \zeta) g(\zeta) \frac{d\zeta}{i\zeta} \qquad (|z| < \rho)$$

minimizes the Dirichlet integral.

Proof: According to Theorem 9 of the previous section $u(z)$ takes on the given boundary values $g(\zeta)$. What precludes direct application of Theorem 4 is the fact that nothing is known concerning the existence of boundary values of the derivatives u_x, u_y, u_{xx}, u_{yy}. To overcome this difficulty we approximate the function $u(z)$ by polynomials $u_n(z)$ in x and y as follows: Going back to the formulas (6), (7), (8) of the preceding section we put

$$a_n = \frac{1}{\pi} \int_C \zeta^{-n} g(\zeta) \frac{d\zeta}{i\zeta} \quad (n = 0, 1, \ldots), \quad f(z) = \tfrac{1}{2} a_0 + \sum_{n=1}^{\infty} a_n z^n \quad (|z| < \rho),$$

so that

$$u(z) = \frac{f(z) + \overline{f(z)}}{2},$$

and

(5) $\dfrac{1}{\pi} \displaystyle\int_C \zeta^n g(\zeta) \dfrac{d\zeta}{i\zeta} = \dfrac{1}{\pi} \displaystyle\int_C \rho^{2n}\bar\zeta^{-n} g(\zeta) \dfrac{d\zeta}{i\zeta} = \bar a_n \rho^{2n}$ $(n = 0, 1, \ldots).$

We now use the partial sums

$$f_n(z) = \tfrac{1}{2}a_0 + \sum_{k=1}^{n} a_k z^k \qquad (n = 0, 1, \ldots)$$

of the power series $f(z)$ to define the polynomials

$$u_n(z) = \frac{f_n(z) + \overline{f_n(z)}}{2} \qquad (n = 0, 1, \ldots).$$

As a polynomial in z, $f_n(z)$ is regular in the z-plane. As the real part of $f_n(z)$, $u_n(z)$ is harmonic in the z-plane. Further, the sequence $u_n(z)$ converges in the interior of the circle C to $u(z)$, and this convergence is uniform in the interior of every smaller concentric circle.

With the polynomial $u_n(z)$ in place of u we may apply Theorem 3, and note that $\Delta u_n = 0$. For every function v which is continuous and piecewise differentiable on $\mathfrak{B}$ and has a convergent Dirichlet integral we have

(6) $$D[u_n, v; \mathfrak{B}] = \int_C v(u_{nx}\, dy - u_{ny}\, dx).$$

Now let $v = w - u_n$, where w is any function which is continuous and piecewise differentiable on $\mathfrak{B}$, has a convergent Dirichlet integral and takes on the boundary values $g(\zeta)$. We must show that

$$\int_C v(u_{nx}\, dy - u_{ny}\, dx) = 0,$$

in spite of the fact that the function v need not vanish identically on the boundary. If

$$v_n(z) = \frac{f_n(z) - \overline{f_n(z)}}{2i} \qquad (n = 0, 1, \ldots)$$

is the harmonic conjugate of $u_n(z)$ then

$$u_{nx}\, dy - u_{ny}\, dx = v_{ny}\, dy + v_{nx}\, dx = dv_n,$$

and, in particular, we have on C

$$4v(u_{nx}\, dy - u_{ny}\, dx) = \left\{ 2g(\zeta) - a_0 - \sum_{k=1}^{n}(a_k \zeta^k + \bar a_k \bar\zeta^k) \right\} \sum_{k=1}^{n} k(a_k \zeta^k + \bar a_k \bar\zeta^k) \frac{d\zeta}{i\zeta}.$$

Using this relation and (5) we obtain

$$\int_C v(u_{nx}\,dy - u_{ny}\,dx) = \pi \sum_{k=1}^{n} k a_k \bar{a}_k \rho^{2k} - \pi \sum_{k=1}^{n} k a_k \bar{a}_k \rho^{2k} = 0.$$

Hence (6) becomes

$$D[u_n, v; \mathfrak{B}] = 0,$$

and, in view of Theorem 2, we obtain

$$D[w; \mathfrak{B}] = D[u_n + v; \mathfrak{B}] = D[u_n; \mathfrak{B}] + D[v; \mathfrak{B}] \geq D[u_n; \mathfrak{B}].$$

It remains to show that the latter inequality holds with u in place of u_n. To this end we form the Dirichlet integral $D_r[u_n]$ of u_n over the closed disk $|z| \leq r, 0 < r < \rho$. Then

(7)
$$\iint\limits_{|z| \leq r} (u_{nx}^2 + u_{ny}^2)\,dx\,dy = D_r[u_n]$$

$$\leq D[u_n; \mathfrak{B}] \leq D[w; \mathfrak{B}].$$

In the disk under consideration the sequence of functions u_n converges uniformly to u. In view of Theorem 10 of the preceding section this implies the uniform convergence of the derivatives u_{nx} and u_{ny} to u_x and u_y. Now we can go over in (7) to the limit $n \to \infty$ under the integral sign and obtain the inequality

$$D_r[u] \leq D[w; \mathfrak{B}],$$

and, therefore also for $r \to \rho$, the inequality

(8)
$$D[u; \mathfrak{B}] \leq D[w; \mathfrak{B}].$$

This proves the assertion of the theorem provided we show that we can have equality in (8) only if $w = u$. To prove this we form, as in the beginning of the proof of Theorem 5, the function $u + \varepsilon(w - u)$, and apply to this function in place of w the inequality (8). In this way we obtain

$$D[u, w - u; \mathfrak{B}] = 0, \qquad D[w; \mathfrak{B}] = D[u; \mathfrak{B}] + D[w - u; \mathfrak{B}].$$

If $D[w; \mathfrak{B}] = D[u; \mathfrak{B}]$, then it follows that $D[w - u; \mathfrak{B}] = 0$ and from this, as in the proof of Theorem 4, we obtain $w = u$.

7. Preliminaries for the mapping theorem

In Sections 8 and 9 we shall be concerned with the proof of a generalization of the Riemann mapping theorem. This theorem, for which Riemann gave no rigorous proof, states that every schlicht simply connected domain of the number sphere with a minimum of two boundary points can be mapped conformally onto the interior of the unit circle. It is possible that this theorem was suggested to Riemann by physical considerations of the type later used very frequently for heuristic purposes by Klein. Just such an argument follows.

We think of the Riemann number sphere as being a uniform electrical conductor. We let an electrical current enter at one point of the sphere and exit at another and suppose a steady flow established. We think of the two points as coming together at the origin. We assume that the resulting dipole has the orientation of the positive x-axis. Under certain physical assumptions (which we shall not specify here) the flow has an electrical potential u which, for $z \neq 0$, is a harmonic function. In the present case

$$u + iv = \frac{1}{z} = \frac{1}{x + iy},$$

where v is the harmonic conjugate of u. It follows that

$$u = \frac{x}{x^2 + y^2}, \qquad v = \frac{-y}{x^2 + y^2}.$$

The equipotential lines $u = c$ with constant real c yield the family of circles through the origin tangent to the y-axis, and the flow lines $v = c$ are their orthogonal trajectories, that is, the family of circles through the origin tangent to the x-axis. Beginning at the origin, the flow moves first in the direction of increasing x and then returns to the origin. For a given $c \neq 0$ the circle has radius $1/2|c|$, and for $c = 0$ we obtain the axes.

In place of the sphere we now take a bounded plane domain $\mathfrak{G}$ bounded by a simple closed curve C and containing the origin. We again introduce at the origin an electrical dipole whose direction is that of the positive x-axis. If the domain together with its boundary is a uniform electrical conductor, then a steady flow is established whose electrical potential u is harmonic on $\mathfrak{G}$ for $z \neq 0$. If u is the real part of the analytic function $f(z) = u + iv$, then the function $f(z) - z^{-1}$ turns out to be regular in $\mathfrak{G}$, so that $f(z)$ has a simple pole at the origin. For large values of $|c|$ the curves $u = c$ and $v = c$ behave roughly like the corresponding circles in the previous example. Also, it is reasonable to expect that all but one of the flow lines $v = c$, c an arbitrary constant, are closed, and that the exceptional flow line $v = c_0$ goes from the origin to a boundary point ζ_1, where it forks, and then returns from another boundary point ζ_2 to the origin (Figure 57).

We now consider the mapping of $\mathfrak{G}$ onto the (u, v)-plane effected by the function $f(z) = u + iv = w$. To the flow lines $v = c$ in $\mathfrak{G}$ there correspond in the w-plane lines parallel to the real axis. Since the constant c can take on all real values, the whole w-plane is covered. The point $z = 0$ is taken to the point $w = \infty$. The image of the flow line $v = c_0$ is not a simple line parallel to the real axis, because a certain segment L on this parallel is the image of the

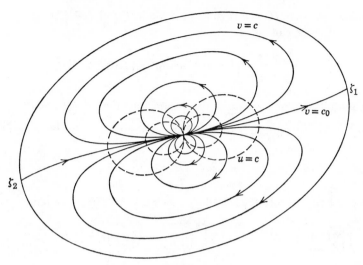

Figure 57

two parts of the boundary C determined by the points ζ_1 and ζ_2 and, as such, is covered twice (Figure 58). In other words, the function $w = f(z)$ maps the region $\mathfrak{G}$ conformally onto the w-sphere slit along L. By means of a fractional linear transformation it is possible to carry the end points of the slit into 0, ∞ and the slit into the negative real half-axis. The plane slit in this manner can be mapped by means of the function $\sqrt{w}$ onto the right half plane, and the latter can be mapped by another fractional linear transformation onto the interior of the unit circle.

The above considerations suggest that the crucial point in the proof of the Riemann mapping theorem is the existence of a function which for $z \neq 0$ is harmonic in $\mathfrak{G}$, and at $z = 0$ has the same singularity as the function $x/(x^2 + y^2)$. It turns out that such a function can be obtained as the solution of a certain extremum problem which is closely related to the extremum problem in the preceding section, and that this method is applicable even to arbitrary Riemann regions. We now prove a few additional preliminary results.

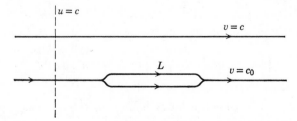

Figure 58

Theorem 1:   Let $f(z)$ be regular analytic in the closed disk $|z| \leq \rho$, then we have for the mean value

$$\frac{1}{\pi \rho^2} \iint_{|z| \leq \rho} |f(z)|^2 \, dx \, dy \geq |f(0)|^2,$$

with equality only in the case of a constant function.

Proof: We introduce polar coordinates by putting $z = re^{i\varphi}$ and make use of the power series

$$f(z) = \sum_{n=0}^{\infty} c_n z^n,$$

which converges absolutely for $|z| \leq \rho$. Here $c_0 = f(0)$. We have

$$\iint_{|z| \leq \rho} |f(z)|^2 \, dx \, dy = \sum_{k,l=0}^{\infty} c_k \bar{c}_l \int_0^{\rho} \left(\int_0^{2\pi} e^{i(k-l)\varphi} \, d\varphi \right) r^{k+l+1} \, dr$$

$$= 2\pi \sum_{n=0}^{\infty} |c_n|^2 \frac{\rho^{2n+2}}{2n+2} \geq \pi |c_0|^2 \rho^2 = \pi \rho^2 |f(0)|^2.$$

Equality holds if and only if $c_n = 0$, for $n = 1, 2, \ldots$, and then $f(z)$ is the constant c_0. The theorem just proved carries over without any difficulty to an arbitrary closed disk $|z - a| \leq \rho$.

Theorem 2:   Let u be harmonic in the interior of a region $\mathfrak{B}$ and have a convergent Dirichlet integral

$$D[u; \mathfrak{B}] = \delta.$$

If the distance of a point z_0 of the region from the boundary is at least equal to a given positive number q, then we have at z_0 the inequality

$$u_x^2 + u_y^2 \leq \frac{\delta}{\pi q^2}.$$

Proof: Let v be the conjugate harmonic function of u and let $u + iv = g(z)$. Then we have for the derivative

$$f(z) = g'(z) = u_x + i v_x = u_x - i u_y,$$

and so, for $0 < \rho < q$, the inequality

$$(1) \qquad \iint_{|z-z_0| \leq \rho} |f(z)|^2 \, dx \, dy = \iint_{|z-z_0| \leq \rho} (u_x^2 + u_y^2) \, dx \, dy \leq \delta.$$

Using Theorem 1 to obtain a lower bound for the first integral in (1) and letting ρ go to q we obtain the required result.

Theorem 3: Let $u_n(z)$ $(n = 1, 2, \ldots)$ be a sequence of functions harmonic in a Jordan domain $\mathfrak{G}$ and having Dirichlet integrals. Further, let

$$\lim_{n \to \infty} D[u_n; \mathfrak{G}] = 0.$$

If the sequence of functional values $u_n(z_0)$ converges to 0 at a point z_0 of $\mathfrak{G}$, then $u_n(z) \to 0$ in all of $\mathfrak{G}$, and this convergence is uniform on every region contained in $\mathfrak{G}$.

Proof: Let $\mathfrak{K}$ be a closed disk of radius ρ containing the point z_0 whose distance from the boundary of the region is q. Let $D[u_n; \mathfrak{G}] = \delta_n$ $(n = 1, 2, \ldots)$. In view of Theorem 2 the following estimate holds on $\mathfrak{K}$:

$$u_{nx}^2 + u_{ny}^2 \le \frac{\delta_n}{\pi q^2} \qquad (n = 1, 2, \ldots).$$

If z is another point of $\mathfrak{K}$, then, by integrating along the line from z_0 to z and applying the Schwarz inequality, we have

$$(2) \qquad \left| \int_{z_0}^{z} (u_{nx}\, dx + u_{ny}\, dy) \right| \le \frac{2\rho}{q} \sqrt{\frac{\delta_n}{\pi}}.$$

On the other hand,

$$(3) \qquad \int_{z_0}^{z} (u_{nx}\, dx + u_{ny}\, dy) = u_n(z) - u_n(z_0).$$

It follows that

$$u_n(z) - u_n(z_0) \to 0, \qquad u_n(z) \to 0 \qquad (n \to \infty)$$

uniformly on $\mathfrak{K}$.

If now z_1 is any point in $\mathfrak{G}$, then z_0 can be joined to z_1 by a curve which can be covered by finitely many closed disks lying in $\mathfrak{G}$. Repeated application of the above argument shows that $u_n(z_1) \to 0$, and hence again by the above argument, $u_n(z) \to 0$ uniformly in any closed disk lying in $\mathfrak{G}$. Since any region $\mathfrak{B}$ lying in $\mathfrak{G}$ can be covered by finitely many closed disks in $\mathfrak{G}$, $u_n(z) \to 0$ uniformly in $\mathfrak{B}$.

Theorem 4: Let $u_n(z)$ $(n = 1, 2, \ldots)$ be a sequence of functions harmonic in a Jordan domain $\mathfrak{G}$ and having Dirichlet integrals. Further, let

$$\lim_{n, m \to \infty} D[u_n - u_m; \mathfrak{G}] = 0.$$

If the sequence of functional values $u_n(z_0)$ converges at a point z_0 of $\mathfrak{G}$, then $u_n(z)$ converges at all points z in $\mathfrak{G}$ to a limit function harmonic in $\mathfrak{G}$, and this convergence is uniform on every region in $\mathfrak{G}$.

Proof: Let ε be an arbitrary given positive number. By assumption there exists a natural number $N = N(\varepsilon)$ such that

$$D[u_n - u_m; \mathfrak{G}] < \varepsilon \qquad (n, m > N).$$

Now we argue as in the proof of the preceding theorem. If $\mathfrak{K}$ is again a closed disk in $\mathfrak{G}$ with radius ρ whose distance from the boundary of the region $\mathfrak{G}$ is q, and if $\mathfrak{K}$ contains the points z and z_0, then, in analogy to (2) and (3), we have the estimate

$$|(u_n(z) - u_m(z)) - (u_n(z_0) - u_m(z_0))| < \frac{2\rho}{q}\sqrt{\frac{\varepsilon}{\pi}} \qquad (n, m > N).$$

In view of the convergence of the sequence $u_n(z_0)$, this implies the uniform convergence of $u_n(z)$ on $\mathfrak{K}$ to a limit function $u(z)$. By Harnack's theorem (Theorem 10, Section 5) $u(z)$ is harmonic in the interior of $\mathfrak{K}$.

Now the theorem follows by an argument similar to that used at the end of Theorem 3.

Next we introduce the *Schwarz reflection principle* for harmonic functions.

Theorem 5:   Let $\mathfrak{B}$ be a region which is divided by an arc L of the circle K into two subregions $\mathfrak{A}$ and $\mathfrak{A}^*$ which are reflections of each other in K. Let $u(z)$ be harmonic in the interior of $\mathfrak{A}$ and vanish on L. Then $u(z)$ can be harmonically continued across L into the full interior of $\mathfrak{B}$.

Proof: If z^* is the image in $\mathfrak{A}^*$ of z in $\mathfrak{A}$ under reflection in K, then we put (Figure 59)

$$u(z^*) = -u(z).$$

This definition is meaningful on L since there $z^* = z$ and $u(z) = 0$. We see that $u(z)$ is defined and continuous in the interior of $\mathfrak{B}$, and it remains to show that it is harmonic at the points of $\mathfrak{A}^*$. To simplify the proof we first carry the circle K into the real axis by means of a suitable fractional linear transformation in z. Then L goes into a real interval and $\mathfrak{A}$ into a subregion of the upper half plane. Since our transformation preserves mirror images, it suffices, in view of Theorem 3 in Section 5, to prove the remainder of our assertion for the simplified case in which $z^* = \bar{z}$.

Every reflection effects a conformal mapping and, in view of the fact that the harmonicity of u implies the harmonicity of $-u$, $u(z)$ is seen to be harmonic in the interior of $\mathfrak{A}^*$. It remains to show that $u(z)$ satisfies the potential equation in the interior of the interval L. We show this to be the case for a definite point which we may take, without restriction of generality, to be the point $z = 0$, and we choose the radius ρ of the closed disk $|z| \leq \rho$ so small that it lies entirely inside $\mathfrak{B}$. By means of the Poisson integral we define

$$w(z) = \frac{1}{2\pi}\int_{-\pi}^{\pi} P(z, \zeta)u(\zeta)\,d\zeta \qquad (\zeta = \rho e^{i\varphi}, |z| < \rho).$$

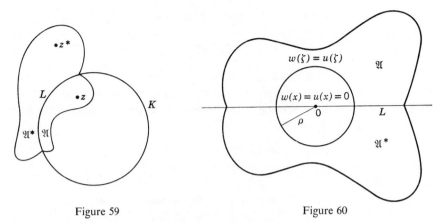

Figure 59   Figure 60

Since

$$u(\bar{\zeta}) = -u(\zeta), \qquad P(z, \bar{\zeta}) = \frac{\bar{\zeta}\zeta - z\bar{z}}{(\bar{\zeta} - z)(\zeta - \bar{z})} = P(\bar{z}, \zeta),$$

and the transition from ζ to $\bar{\zeta}$ corresponds to the interchanging of φ and $-\varphi$, we have

$$w(\bar{z}) = \frac{1}{2\pi} \int_{-\pi}^{\pi} P(\bar{z}, \zeta)u(\zeta)\, d\varphi = -\frac{1}{2\pi} \int_{-\pi}^{\pi} P(z, \bar{\zeta})u(\bar{\zeta})\, d\varphi$$

$$= -\frac{1}{2\pi} \int_{-\pi}^{\pi} P(z, \zeta)u(\zeta)\, d\varphi = -w(z)$$

for $|z| < \rho$. For real $z = x$ of the interval $-\rho < x < \rho$ we have, in particular, $w(z) = 0$. According to Theorem 9 in Section 5, the function $w(z)$ defined by the Poisson integral is harmonic in the open disk $|z| < \rho$ and takes on the boundary values $u(\zeta)$ at the circumference. The functions $u(z)$ and $w(z)$ are harmonic in the open half disk $|z| < \rho$, $y > 0$, in $\mathfrak{A}$, and have the same boundary values there (Figure 60). By Theorem 8 in Section 5 we have, in the half disk in question, the equality $u(z) = w(z)$. Since $u(\bar{z}) = -u(z)$, $w(\bar{z}) = -w(z)$, it follows that this identity holds throughout the circle $|z| < \rho$. To complete the proof it suffices to note that $w(z)$ is harmonic in $|z| < \rho$.

Theorem 6: Let $u_n(n = 1, 2, \ldots)$ be a sequence of functions harmonic inside a region $\mathfrak{B}$ and having Dirichlet integrals. Let L be a circular arc which is a portion of the boundary of $\mathfrak{B}$, and let each function u_n vanish on L. Further, let

$$\lim_{n \to \infty} D[u_n; \mathfrak{B}] = 0.$$

Then $u_n \to 0$ in the interior of $\mathfrak{B}$, and this convergence is uniform on every region in $\mathfrak{B}$.

Proof: Consider a circle orthogonal to L and so small that of the two regions $\mathfrak{A}$ and $\mathfrak{A}^*$ into which L divides the closed disk $\mathfrak{K}$, one, say $\mathfrak{A}$, belongs entirely to $\mathfrak{B}$ whereas $\mathfrak{A}^*$ and $\mathfrak{B}$ have in common only points of the arc L (Figure 61). Then $\mathfrak{A}^*$ is the image of $\mathfrak{A}$ under reflection in L, and we can apply Theorem 5 to the function $u = u_n$ with $\mathfrak{K}$ in place of $\mathfrak{B}$. It follows that the harmonic continuation of u_n exists throughout the interior of $\mathfrak{K}$. In view of Theorem 1 in Section 6 the Dirichlet integral is invariant under reflection and it follows that

$$D[u_n; \mathfrak{K} \cup \mathfrak{B}] = D[u_n; \mathfrak{A}^*] + D[u_n; \mathfrak{B}]$$
$$= D[u_n; \mathfrak{A}] + D[u_n; \mathfrak{B}] \leq 2D[u_n; \mathfrak{B}] \to 0$$

for $n \to \infty$. To obtain the required result we need only apply Theorem 3 with $\mathfrak{G}$ replaced by the interior of $\mathfrak{K} \cup \mathfrak{B}$ and with z_0 a point of L.

8. Construction of a harmonic function with minimal property

By virtue of its definition, a Riemann region $\mathfrak{R}$ over the z-sphere consists of disks connected in a definite manner. At the branch points the disks are connected in the manner indicated by the local uniformizing variable. *In this section we shall always mean by the term disk a closed disk.* The interior of a disk $\mathfrak{K}$ will be denoted by $\underline{\mathfrak{K}}$.

Theorem 1: Every Riemann region $\mathfrak{R}$ can be covered with countably many disks belonging to $\mathfrak{R}$, so that every point of $\mathfrak{R}$ is in the interior of one of the covering disks.

Proof: First we consider the punctured Riemann region $\mathring{\mathfrak{R}}$ obtained from $\mathfrak{R}$ by omitting the branch points and the points over $z = \infty$. We shall call a point $\mathfrak{z}$ of $\mathring{\mathfrak{R}}$ rational if its projection $z = x + iy$ has rational coordinates x and y. Let $\mathfrak{z}_0$ be a fixed rational point. If $\mathfrak{z}$ is any other rational point we can join $\mathfrak{z}_0$ and $\mathfrak{z}$ by means of a curve on $\mathring{\mathfrak{R}}$ and, therefore also by means of a curve L on $\mathring{\mathfrak{R}}$ with polygonal projection whose successive vertices $\mathfrak{z}_0$, $\mathfrak{z}_1, \ldots, \mathfrak{z}_n = \mathfrak{z}$ are all rational. The point $\mathfrak{z}_0$ and the projections $z_1, \ldots, z_n$ of $\mathfrak{z}_1, \ldots, \mathfrak{z}_n$, in this order, determine L on $\mathring{\mathfrak{R}}$ uniquely and, in particular, the end point $\mathfrak{z}$ of L. On the other hand, let

$$z_k = \frac{q_k + ir_k}{s_k}, \qquad (q_k, r_k, s_k) = 1, \qquad s_k > 0 \qquad (k = 1, \ldots, n).$$

If m is a natural number and

$$n + \sum_{k=1}^{n}(|q_k| + |r_k| + |s_k|) = m,$$

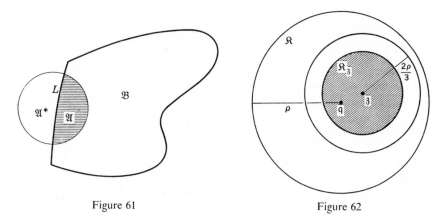

Figure 61   Figure 62

then, clearly, the number of possibilities for n and $z_1, \ldots, z_n$ is finite. It follows that there are only countably many rational points on $\mathfrak{R}$.

For a rational point $\mathfrak{z}$ let $a_\mathfrak{z}$ denote the (possibly infinite) least upper bound of all radii of the projections of the closed disks on $\mathfrak{R}$ with center $\mathfrak{z}$. Among these disks we single out the disk $\mathfrak{R}_\mathfrak{z}$ whose radius $\rho_\mathfrak{z}$ is the smaller one of the two numbers 1 and $\frac{1}{2}a_\mathfrak{z}$. For any point q on $\mathfrak{R}$ let $\mathfrak{R}$ be a disk on $\mathfrak{R}$ with center q and radius ρ. $\mathfrak{R}$ contains a rational point $\mathfrak{z}$ whose distance from q is smaller than either of the numbers 1 and $\frac{1}{3}\rho$, and so the closed disk with center $\mathfrak{z}$ and radius $\frac{2}{3}\rho$ lies completely in $\mathfrak{R}$. It follows that $\frac{2}{3}\rho \leq a_\mathfrak{z}$, $\frac{1}{3}\rho \leq \frac{1}{2}a_\mathfrak{z}$ and, therefore q is in the interior of $\mathfrak{R}_\mathfrak{z}$ (Figure 62). This proves that the countably many disks $\mathfrak{R}_\mathfrak{z}$ cover the punctured Riemann region $\mathring{\mathfrak{R}}$, and every point of $\mathfrak{R}$ belongs to the interior of one such disk.

It remains to prove that the branch points and the points on $\mathfrak{R}$ over $z = \infty$ can be covered with either finitely many or with countably many disks. If q is a point on $\mathfrak{R}$ whose projection is ∞ then we choose a disk $\mathfrak{R}$ on $\mathfrak{R}$ about q corresponding to the local uniformizing parameter, and in it a point $\mathfrak{z}$, not a branch point, with rational projection z. Since prescribing $\mathfrak{z}$ determines the point q of $\mathfrak{R}$ over $z = \infty$ uniquely, it follows from the countability of the rational points that there are at most countably many q which are covered by the corresponding disks $\mathfrak{R}$. In this argument q may be a branch point. Finally, let q be a branch point with finite projection. We consider the circles on $\mathfrak{R}$ about q corresponding to the local uniformizing parameter and define a_q, ρ_q, $\mathfrak{R}_q$ as above with $\mathfrak{z}$ in place of q. If q_1 and q_2 are two different branch points then, as we are about to show, the disks $\mathfrak{R}_{q_1}$ and $\mathfrak{R}_{q_2}$ are disjoint. In fact, let z be the projection of a common point of $\mathfrak{R}_{q_1}$ and $\mathfrak{R}_{q_2}$; let z_k ($k = 1, 2$) be the projection of q_k; and let j_k be the number of sheets connected at q_k. If we put $\rho_{q_k} = \rho_k$, then we have

$$|z_1 - z_2| \geq (2\rho_k)^{j_k} > 2\rho_k^{j_k} \quad (k = 1, 2), \qquad |z_1 - z_2| > \rho_1^{j_1} + \rho_2^{j_2}.$$

On the other hand,

$$|z_1 - z_2| \leq |z - z_1| + |z - z_2| \leq \rho_1^{j_1} + \rho_2^{j_2}.$$

A contradiction. Since every disk $\Re$ contains a rational point $\mathfrak{z}$, it now follows that the number of branch points $\mathfrak{q}$ and of disks $\Re_{\mathfrak{q}}$ which cover them is at most countable. This completes the proof.

Theorem 2: On every Riemann region $\Re$ it is possible to choose finitely many disks $\Re_1, \ldots, \Re_\nu$ or countably many disks $\Re_1, \Re_2, \ldots$ which cover $\Re$ in the sense of Theorem 1 such that for $n = 1, \ldots, \nu - 1$ and $n = 1, 2, \ldots,$ respectively, the union of $\underline{\Re}_1, \ldots, \underline{\Re}_n$ intersects $\underline{\Re}_{n+1}$ without containing it. Also, the number of required disks is finite if and only if $\Re$ is compact.

Proof: In view of Theorem 1 there is a sequence of disks $\mathfrak{H}_1, \mathfrak{H}_2, \ldots$ which cover $\Re$. We choose $\Re_1 = \mathfrak{H}_1$ and proceed by induction. For a given natural number r we determine finitely many disks $\Re_1, \Re_2, \ldots, \Re_s$ with the following properties. For $n = 1, \ldots, s - 1$ the union of $\underline{\Re}_1, \ldots, \underline{\Re}_n$ intersects the open disk $\underline{\Re}_{n+1}$ but does not fully contain it; further, the union of $\underline{\Re}_1, \ldots, \underline{\Re}_s$ covers the open disks $\mathfrak{H}_1, \ldots, \mathfrak{H}_r$. This assumption is trivially satisfied for $r = 1$ with $s = 1$. Suppose, now, that $\underline{\Re}_1, \ldots, \underline{\Re}_s$ do not cover all of $\mathfrak{H}$ and let $\mathfrak{H}_t$ be the first open disk in the sequence $\mathfrak{H}_{r+1}, \mathfrak{H}_{r+2}, \ldots$ which is not contained in the union of $\underline{\Re}_1, \ldots, \underline{\Re}_s$. We join $\mathfrak{H}_r$ and $\mathfrak{H}_t$ by means of a finite chain of circles $\mathfrak{H}_r, \mathfrak{H}_l, \mathfrak{H}_m, \ldots, \mathfrak{H}_t$ of which any two successive ones overlap. We omit each link which is contained in the union of its predecessors and $\underline{\Re}_1, \ldots, \underline{\Re}_s$, and denote the others, without changing their order, by $\underline{\Re}_{s+1}, \underline{\Re}_{s+2}, \ldots, \underline{\Re}_w$. Hence the induction assumption is fulfilled for $t > r$ with w in place of s. Our procedure yields the required sequence $\Re_1, \Re_2, \ldots$. When this sequence breaks off, then, bearing in mind that the individual disks are closed, we see that $\Re$ is compact. If this sequence does not break off, then for $n = 1, 2, \ldots,$ we choose in $\underline{\Re}_n$ a point $\mathfrak{y}_n$ which is not in $\underline{\Re}_1, \ldots, \underline{\Re}_{n-1}$. If the sequence of the $\mathfrak{y}_n$ had a limit point $\mathfrak{y}$ on $\Re$, then $\mathfrak{y}$ would lie in some open disk $\underline{\Re}_n$, and $\mathfrak{y}_{n+1}, \mathfrak{y}_{n+2}, \ldots$ would lie outside that disk, which is impossible. This shows that in this case $\Re$ is not compact. The proof is complete.

Arguing as above we can easily show that every compact set $\mathfrak{M}$ on $\Re$ can be covered by a finite number of the $\underline{\Re}_n$. To this end we form the intersection $\mathfrak{D}_n$ $(n = 1, 2, \ldots)$ of $\mathfrak{M}$ with the union of $\Re_1, \ldots, \Re_n$. If the assertion were false we could find in $\mathfrak{M}$ for $n = 1, 2, \ldots$ a point $\mathfrak{y}_n$ not in $\mathfrak{D}_n$, and for every limit point $\mathfrak{y}$ of the $\mathfrak{y}_n$ on $\Re$ we would obtain the above contradiction.

For what follows it is convenient to choose the disks $\Re_1, \Re_2, \ldots$ in a special way. Let $\mathfrak{z}_0$ be a point on $\Re$ which is not a branch point or a point at infinity. A translation justifies the assumption that $\mathfrak{z}_0$ lies over $z = 0$. Let

$\Re_1$ be a disk on $\Re$ about $\mathfrak{z}_0$, and let $\Re_0$ be a smaller concentric disk. Then $\Re - \Re_0$ is also a Riemann region and, by Theorem 1, can be covered with disks $\mathfrak{H}_2, \mathfrak{H}_3, \ldots$. Together with $\mathfrak{H}_1 = \Re_1$ these provide a covering of $\Re$. The covering $\Re_1, \Re_2, \ldots$ constructed in accordance with the procedure in the proof of Theorem 2 has the property, to be used in the sequel, that $\Re_0$ is interior to $\Re_1$ and exterior to the remaining circles $\Re_2, \Re_3, \ldots$. Let $\mathfrak{B}_n$ denote the union of $\Re_1, \ldots, \Re_n$, and $\underline{\mathfrak{B}}_n$ the union of the open disks $\underline{\Re}_1, \ldots, \underline{\Re}_n$. $\underline{\mathfrak{B}}_n$ is a connected Riemann subregion of $\Re$ whose closure is $\mathfrak{B}_n$. If $\mathfrak{M}$ is a set on $\Re$ containing $\mathfrak{z}_0$, then $\dot{\mathfrak{M}}$ will denote the set $\mathfrak{M}$ with $\mathfrak{z}_0$ omitted.

A real valued function $w = w\,(\mathfrak{z})$ defined on $\Re$ is said to be *continuous* or *differentiable* if it has these properties everywhere when regarded as a function of the real and imaginary parts of the local uniformizing parameter, and this definition carries over without modifications to subdomains of $\Re$. A *harmonic function* is defined in a similar manner; here it should be pointed out that in view of the invariance under conformal mapping established in Theorem 3, Section 5, no difficulties arise even in the vicinity of points at infinity and branch points. Let $C_0, C_1, \ldots$ be the boundaries of $\Re_0, \Re_1, \ldots$. We shall call a continuous function on $\Re$ piecewise differentiable if it has continuous first order partial derivatives with the possible exception of points of finitely many circumferences C_n on $\Re$. If h_1 and h_2 are such functions then so is $\lambda_1 h_1 + \lambda_2 h_2$, where λ_1 and λ_2 are arbitrary real constants. For $n = 1, 2, \ldots$ it is possible to divide the region $\mathfrak{B}_n$ consisting of $\dot{\Re}_1$, $\Re_2, \ldots, \Re_n$ into finitely many nonoverlapping subregions bounded by circular arcs over the z-plane inside of which the function h is continuously differentiable. As one of these subregions we can take $\dot{\Re}_0$. If the Dirichlet integral of h converges in all of these subregions with the exception of $\Re_0$, then the sum of the integrals in question exists for every fixed n and is denoted by $D(h; \mathfrak{B}_n - \Re_0)$. It should be noted that the definition of piecewise differentiability refers to a fixed covering of $\Re$ with disks $\Re_0, \Re_1, \ldots$ of the described kind. In connection with the functions h in $\dot{\Re}_0$ to be studied in the sequel we make the following special assumption concerning their behavior when $\mathfrak{z}$ approaches $\mathfrak{z}_0$. Let a be the radius of $\Re_0$. The real part of the function $z^{-1} + a^{-2}z$ regular in $\Re$ for $z \neq 0$ is

$$q(\mathfrak{z}) = q(x, y) = \frac{x}{x^2 + y^2} + \frac{x}{a^2} .$$

We require that for every sequence of points $\mathfrak{z}$ in $\dot{\Re}_0$ converging to $\mathfrak{z}_0$ the difference $h(\mathfrak{z}) - q(\mathfrak{z})$ converges to a limit, that is, it is continuous at $\mathfrak{z}_0$. We also require the existence of the integral $D\,[h - q; \dot{\Re}_0]$ which we denote for brevity by $D\,[h - q; \Re_0]$. In view of Theorem 2 of Section 6 these requirements hold if they hold for the simpler function $x(x^2 + y^2)^{-1}$ in place of q.

Nevertheless, the stated form of the convergence condition turns out to be convenient. We now define $\hat{h} = h - q$ for $\mathfrak{z}$ in $\mathfrak{K}_0$ and $\hat{h} = h$ in $\mathfrak{R} - \mathfrak{K}_0$ so that $\hat{h}$ in $\mathfrak{R}$ is discontinuous on C_0. Because of the singularity at $\mathfrak{z}_0$ the Dirichlet integral of h over $\mathfrak{B}_n$ fails to exist and in its place we form the modified expression

$$D[h; \mathfrak{B}_n - \mathfrak{K}_0] + D[h - q; \mathfrak{K}_0] = D[\hat{h}; \mathfrak{B}_n].$$

It is clear that for $n = 1, 2, \ldots$ these values form a monotonic increasing sequence of nonnegative numbers. If the sequence is bounded, and therefore has a limit, then we define

$$D[\hat{h}; \mathfrak{R}] = \lim_{n \to \infty} D[\hat{h}; \mathfrak{B}_n],$$

and refer to this number as the *normalized Dirichlet integral* of h over $\mathfrak{R}$.

We agree to call h a *comparison function* provided it has the above three properties. Such a function h then is continuous and piecewise differentiable in $\mathfrak{R}$, the function $h - q$ is continuous at $\mathfrak{z}_0$, and the Dirichlet integral $D[\hat{h}; \mathfrak{R}]$ converges. For our purposes it is essential that h has the prescribed singularity at $\mathfrak{z}_0$, and that is why we had to modify the definition of the Dirichlet integral in the indicated manner; indeed, but for this modification the integral would always diverge. It must nevertheless be shown that there are comparison functions. To this end we introduce on $\mathfrak{K}_1$ polar coordinates $x = r \cos \varphi, y = r \sin \varphi \ (0 \leq r \leq b)$, where b is the radius of $\mathfrak{K}_1$, and again a is the radius of $\mathfrak{K}_0$, and put

$$h = \frac{x}{x^2 + y^2} = \frac{\cos \varphi}{r},$$

$$h = \frac{\cos \varphi}{r} \frac{b - r}{b - a}, \qquad h = 0,$$

for $\mathfrak{z}$ in $\mathfrak{K}_0$, $\mathfrak{K}_1 - \mathfrak{K}_0$, and $\mathfrak{R} - \mathfrak{K}_1$, respectively (Figure 63). It is easy to see that h as defined has all the required properties.

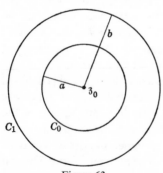

Figure 63

Since $D[\hat{h}; \mathfrak{R}] \geq 0$, it follows that the greatest lower bound of the Dirichlet integrals of all comparison functions h is a nonnegative number μ. Hence there exists a sequence of comparison functions $h_n \ (n = 1, 2, \ldots)$ with $D[\hat{h}_n; \mathfrak{R}] \to \mu$, and every such sequence is called a *minimizing sequence*. We shall show, using suitable minimizing sequences, that the greatest lower bound μ is actually attained as the minimum value of $D[\hat{h}; \mathfrak{R}]$ for a comparison function, and that this extremum function is uniquely determined

apart from an additive constant. This constant could be fixed by the requirement $\hat{h}(\mathfrak{z}_0) = 0$, but it is not expedient to impose this condition a priori on the
comparison functions.

Theorem 3: There exists a comparison function u with

$$D[\hat{u}; \mathfrak{R}] = \mu,$$

and u is harmonic in $\dot{\mathfrak{R}}$.

Proof: The proof is in five steps.

I. Let g and h be two comparison functions. Then their difference

$$f = g - h = (g - q) - (h - q)$$

is continuous in $\mathfrak{R}_0$ and, in view of the existence of $D[g - q; \mathfrak{R}_0]$ and
$D[h - q; \mathfrak{R}_0]$, Theorem 2 in Section 6 insures the convergence of the
integral $D[f; \mathfrak{R}_0]$. In this way we establish also the existence of the two limits

$$\lim_{n \to \infty} D[f; \mathfrak{B}_n] = D[f; \mathfrak{R}], \qquad \lim_{n \to \infty} D[f, \hat{h}; \mathfrak{B}_n] = D[f, \hat{h}; \mathfrak{R}]$$

and the validity of the formula

$$D[\lambda f + \hat{h}; \mathfrak{R}] = \lambda^2 D[f; \mathfrak{R}] + 2\lambda D[f, \hat{h}; \mathfrak{R}] + D[\hat{h}; \mathfrak{R}]$$

for every real constant λ. It follows, in particular, that $\lambda f + \hat{h}$ is a comparison
function. If we put

$$D[f; \mathfrak{R}] = \alpha, \qquad D[f, \hat{h}; \mathfrak{R}] = \beta, \qquad D[\hat{h}; \mathfrak{R}] - \mu = \gamma,$$

then we have

$$\alpha\lambda^2 + 2\beta\lambda + \gamma = D[\lambda f + \hat{h}; \mathfrak{R}] - \mu \geq 0$$

for all real λ, and therefore

$$\beta^2 \leq \alpha\gamma, \; |D[f, \hat{h}; \mathfrak{R}]| \leq \sqrt{D[f; \mathfrak{R}](D[\hat{h}; \mathfrak{R}] - \mu)}.$$

This inequality is also valid with g in place of h. In view of

$$D[f; \mathfrak{R}] = D[f, \hat{g}; \mathfrak{R}] - D[f, \hat{h}; \mathfrak{R}] \leq |D[f, \hat{g}; \mathfrak{R}]| + |D[f, \hat{h}; \mathfrak{R}]|,$$

we obtain

$$\sqrt{D[g - h; \mathfrak{R}]} \leq \sqrt{D[\hat{g}; \mathfrak{R}] - \mu} + \sqrt{D[\hat{h}; \mathfrak{R}] - \mu}.$$

Replacing g and h by two members of a minimizing sequence h_n ($n =
1, 2, \ldots$), we get

(1) $$\lim_{m, n \to \infty} D[h_m - h_n; \mathfrak{R}] = 0.$$

This relation does not prove that the sequence of functions h_n converges
anywhere on $\mathfrak{R}$. We shall, however, find it possible to prove convergence
for minimizing sequences constructed in a special way.

II. We now propose to construct a minimizing sequence of functions harmonic in $\underline{\mathfrak{R}}_1$. Let h be a comparison function. Since the function $h - q$ is continuous on $\mathfrak{R}_1$ and, in particular, on the boundary C_1, it follows, by Theorems 8 and 9 in Section 5, that there is exactly one function which is harmonic in $\underline{\mathfrak{R}}_1$, continuous on $\mathfrak{R}_1$ and coincident with $h - q$ on C_1. We denote this function in $\mathfrak{R}_1$ by $g - q$. If we put $g = h$ in $\mathfrak{R} - \mathfrak{R}_1$ then g is continuous in all of $\mathfrak{R}$ and is admissible as a comparison function.

We claim that if we replace h by g then the normalized Dirichlet integral over $\mathfrak{R}$ does not increase. Clearly we must show that

(2) $D[g; \mathfrak{R}_1 - \mathfrak{R}_0] + D[g - q; \mathfrak{R}_0] \leq D[h; \mathfrak{R}_1 - \mathfrak{R}_0] + D[h - q; \mathfrak{R}_0].$

By Theorem 6 of Section 6, the harmonic function $g - q$ with boundary values on C_1 given by $h - q$ minimizes the Dirichlet integral over $\mathfrak{R}_1$. The inequality

$$D[g - q; \mathfrak{R}_1] \leq D[h - q; \mathfrak{R}_1]$$

implies

$$D[g; \mathfrak{R}_1 - \mathfrak{R}_0] - 2D[q, g; \mathfrak{R}_1 - \mathfrak{R}_0] + D[g - q; \mathfrak{R}_0]$$
$$\leq D[h; \mathfrak{R}_1 - \mathfrak{R}_0] - 2D[q, h; \mathfrak{R}_1 - \mathfrak{R}_0] + D[h - q; \mathfrak{R}_0].$$

This surely yields the formula (2) provided that we can prove the equality

$$D[q, g - h; \mathfrak{R}_1 - \mathfrak{R}_0] = 0.$$

To this end we make use of Green's identity of Theorem 3 in Section 6 with $\mathfrak{B}$ the ring $\mathfrak{R}_1 - \mathfrak{R}_0$ bounded by the circles C_1 and C_0, $u = q$ and $v = g - h$. Since q has derivatives of all orders in the interior of $\mathfrak{B}$ and these derivatives have boundary values, and since $g - h$ is continuous and piecewise differentiable and has a convergent Dirichlet integral over $\mathfrak{B}$, the assumptions of that theorem are satisfied and we have

$$D[q, g - h; \mathfrak{R}_1 - \mathfrak{R}_0] + \iint\limits_{\mathfrak{R}_1 - \mathfrak{R}_0} (\Delta q)(g - h) \, dx \, dy$$

$$= \int_{C_1} (g - h)(q_x \, dy - q_y \, dx) - \int_{C_0} (g - h)(q_x \, dy - q_y \, dx).$$

Here the second double integral vanishes because q is harmonic, and the first line integral vanishes because g and h have the same boundary values on C_1. Also, the second line integral vanishes, since in polar coordinates

$$x = r \cos \varphi, \qquad y = r \sin \varphi, \qquad q = (r^{-1} + ra^{-2}) \cos \varphi,$$

(3) $q_x y_\varphi - q_y x_\varphi = q_x r x_r + q_y r y_r = r q_r = (ra^{-2} - r^{-1}) \cos \varphi = 0 \, (r = a).$

This means that we actually have

$$D[\hat{g}; \mathfrak{R}] \leq D[\hat{h}; \mathfrak{R}].$$

The relation (3) also makes it clear why the additional summand xa^{-2} appears in q.

If we replace each member h of a minimizing sequence by the corresponding function g, then we again obtain the members of a minimizing sequence whose normalized Dirichlet integrals converge to μ at least as well as the earlier sequence. We may therefore assume that for all members h_n ($n = 1, 2, \ldots$) of the minimizing sequence under consideration, the differences $h_n - q$ are harmonic in $\underline{\mathfrak{R}}_1$ and have boundary values throughout C_1. Since, trivially, the value of the Dirichlet integral is not affected by adding a real constant to the function, we may assume that all the $h_n - q$ vanish at $\mathfrak{z}_0$. We shall refer to such a minimum sequence as a *smoothed minimizing sequence*. It is now apparent that the reason for not requiring the comparison functions to have continuous derivatives throughout $\mathfrak{R}$ is that generally, upon transition to a smoothed minimizing sequence, this requirement is violated on C_1.

III.   Next we show that in the case of a smoothed minimizing sequence the differences $h_n - q$ ($n = 1, 2, \ldots$) converge in $\underline{\mathfrak{R}}_1$ to a harmonic function u_0. In view of (1) we certainly have

$$\lim_{m,n \to \infty} D[(h_m - q) - (h_n - q); \mathfrak{R}_1] = 0.$$

Since the functions $h_n - q$ ($n = 1, 2, \ldots$) vanish at $\mathfrak{z}_0$ and are harmonic in $\underline{\mathfrak{R}}_1$ it follows, by Theorem 4, Section 7, that

$$\lim_{n \to \infty} (h_n - q) = u_0$$

in $\underline{\mathfrak{R}}_1$, that the convergence is uniform in the interior of every smaller concentric circle and that the limit function u_0 is harmonic there. Then the function $u = u_0 + q$ is harmonic in $\underline{\mathfrak{R}}_1$, and has the same singularity at $\mathfrak{z}_0$ as q, in the sense that the difference $u - q$ is also harmonic at $\mathfrak{z}_0$ and has there the value 0.

IV.   In general, it is not true that the smoothed minimizing sequence h_n converges in $\mathfrak{R}$ outside $\dot{\mathfrak{R}}_1$. What is true is that the function u constructed in the previous section and harmonic in $\underline{\mathfrak{R}}_1$ admits of unique harmonic continuation to all the points in $\mathfrak{R}$. The proof is by induction. Let t be a given natural number. We assume the existence of a smoothed minimizing sequence h_{tn} ($n = 1, 2, \ldots$) which converges uniformly in every region in $\mathfrak{B}_t$ to a function u harmonic in the interior of $\mathfrak{B}_t$, and which is just the harmonic continuation of the function already constructed on $\dot{\mathfrak{R}}_1$. Further, let h_{tn} ($n = 1, 2, \ldots$) be piecewise harmonic on $\mathfrak{B}_t$, that is, h_{tn} is to be harmonic at all points of $\mathfrak{B}_t$ which do not belong to any of the circles $C_1, \ldots, C_t$. We shall call a minimum sequence with all these properties a *t times smoothed minimizing sequence*; note that we have already constructed such a minimizing sequence for $t = 1$. Given a t times smoothed minimizing sequence we construct a $(t + 1)$ times smoothed sequence in the following manner.

The disk $\Re_{t+1}$ lies outside $\Re_0$ and intersects $\mathfrak{B}_t$. Using the local uniformizing parameter we form the Poisson integral and determine the function j_{tn} harmonic in $\underline{\Re}_{t+1}$ which has on all of C_{t+1} the boundary values given by h_{tn}. j_{tn} is defined on $\Re$ outside $\Re_{t+1}$ by $j_{tn} = h_{tn}$ $(n = 1, 2, \ldots)$. By Theorem 6 of Section 6 we again have the inequalities

$$D[j_{tn}; \Re_{t+1}] \leq D[h_{tn}; \Re_{t+1}], \qquad D[j_{tn}; \Re] \leq D[h_{tn}; \Re],$$

so that the j_{tn} $(n = 1, 2, \ldots)$ also form a minimizing sequence consisting of comparison functions. We now show that these functions are $(t + 1)$ times smoothed. Since h_{tn} is, by assumption, piecewise harmonic on $\dot{\mathfrak{B}}_t$, it follows that j_{tn} is piecewise harmonic on $\underline{\mathfrak{B}}_{t+1}$. Now we choose in the intersection of $\Re_{t+1}$ and $\mathfrak{B}_t$ a region $\mathfrak{B}$ which abuts on C_{t+1} along an arc L, but does not meet any of the circles $C_1, \ldots, C_t$, and apply Theorem 6 of Section 7 to the sequence of differences

$$u_n = u_{tn} = h_{tn} - j_{tn} \qquad (n = 1, 2, \ldots)$$

(Figure 64). Since the combined sequence h_{tn}, j_{tn} $(n = 1, 2, \ldots)$ is also a minimizing sequence, (1) implies the relations

$$\lim_{n \to \infty} D[u_n; \Re] = 0, \qquad \lim_{n \to \infty} D[u_n; \mathfrak{B}] = 0,$$

which shows that all the assumptions of that theorem hold. It follows that

$$\lim_{n \to \infty} u_{tn} = 0$$

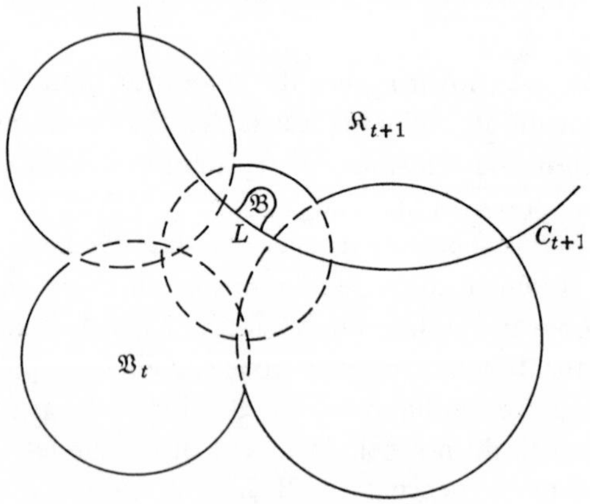

Figure 64

in the interior of $\mathfrak{B}$. In view of the induction assumption we also have there the relation

$$\lim_{n \to \infty} h_{tn} = u,$$

and so the sequence j_{tn} $(n = 1, 2, \ldots)$ converges in the interior of $\mathfrak{B}$ to the function u. On the other hand, we see, as in III, that j_{tn} converges uniformly to a harmonic function in the interior of every smaller circle concentric with C_{t+1}. Since that limit function coincides with u in the interior of $\mathfrak{B}$, it must be the harmonic continuation of u to all of the circular region $\mathfrak{K}_{t+1}$. It is clear that this harmonic continuation is independent of the choice of the arc L and the domain $\mathfrak{B}$. To complete the induction argument we need only put $h_{t+1,n} = j_{tn}$ $(n = 1, 2, \ldots)$.

The inductive procedure enables us to define the function u successively on the regions $\mathfrak{K}_1, \mathfrak{K}_2, \mathfrak{K}_3, \ldots$ in a unique manner so that u is harmonic in each of these domains. Since every point $\mathfrak{z}$ of $\dot{\mathfrak{R}}$ lies in one of these circular domains, it follows that the function u is uniquely determined on $\dot{\mathfrak{R}}$ and is harmonic there.

V. It remains to show that the function u minimizes the normalized Dirichlet integral. Since u is harmonic on $\dot{\mathfrak{R}}$, it is surely continuous and piecewise differentiable on $\dot{\mathfrak{R}}$; further, the difference $u - q$ is continuous at $\mathfrak{z}_0$ and has the value 0 there. For the purpose of investigating the normalized Dirichlet integral we diminish the region $\mathfrak{B}_m$ $(m = 1, 2, \ldots)$ by omitting from each of the covering disks $\mathfrak{K}_1, \ldots, \mathfrak{K}_m$ a boundary strip of width ε. Let $\mathfrak{B}_m(\varepsilon)$ denote the union of the resulting disks $\mathfrak{K}_1(\varepsilon), \ldots, \mathfrak{K}_m(\varepsilon)$. In view of III, on every disk smaller than and concentric with $\mathfrak{K}_1$, $u - q$ is the limit of the sequence $h_n - q$ $(n = 1, 2, \ldots)$ which converges uniformly on every such disk; also, the $h_n = h_{1n}$ form a smoothed minimizing sequence. By the inductive procedure in IV, u is the limit of the sequence $h_{t+1,n}$ $(n = 1, 2, \ldots)$ on every disk smaller than and concentric with $\mathfrak{K}_{t+1}$. On every such disk this sequence is a convergent $(t + 1)$ times smoothed minimizing sequence; also, the $h_{t+1,n}$ $(n = 1, 2, \ldots)$ coincide on $\dot{\mathfrak{R}}$ outside $\mathfrak{K}_{t+1}$ with the h_{tn}. We make use of this fact for $t = 1, \ldots, m - 1$. By Theorem 10 in Section 5 we then have

$$\lim_{n \to \infty} D[\hat{h}_{mn}; \mathfrak{B}_m(\varepsilon)] = D[\hat{u}; \mathfrak{B}_m(\varepsilon)].$$

Since h_{mn} is a minimizing sequence, it follows that

$$\lim D[\hat{h}_{mn}; \mathfrak{R}] = \mu.$$

On the other hand,

$$D[\hat{h}_{mn}; \mathfrak{B}_m(\varepsilon)] \leq D[\hat{h}_{mn}; \mathfrak{R}].$$

Therefore

$$D[\hat{u}; \mathfrak{B}_m(\varepsilon)] \leq \mu,$$

where the right-hand side is independent of ε and m. Letting ε go to 0 and then m go to ∞ we obtain

$$D[\hat{u}; \mathfrak{R}] \leq \mu.$$

Since u is a comparison function, the normalized Dirichlet integral must be $\geq \mu$, and our theorem follows.

We shall now show that prescribing $\mathfrak{z}_0$ determines the extremum function u uniquely, that is, u is independent of the choice of the covering disks $\mathfrak{R}_0$, $\mathfrak{R}_1$, $\mathfrak{R}_2$, ... and of the smoothed minimizing sequence used in the construction.

Theorem 4: Let h be an arbitrary function which is continuous and piecewise differentiable on $\mathfrak{R}$ and for which the Dirichlet integral $D[h; \mathfrak{R}]$ exists. A necessary and sufficient condition for a comparison function u to minimize the normalized Dirichlet integral $D[\hat{u}; \mathfrak{R}]$ is that, for all h,

(4) $$D[\hat{u}, h; \mathfrak{R}] = 0.$$

Proof: If u is a comparison function then so is $u + \lambda h$ for every real constant λ. If

$$D[\hat{u}; \mathfrak{R}] = \mu$$

is the minimum, then the inequality

$$0 \leq D[\hat{u} + \lambda h; \mathfrak{R}] - \mu = \lambda(2D[\hat{u}, h; \mathfrak{R}] + \lambda D[h; \mathfrak{R}])$$

again implies the validity of (4). Conversely, if (4) holds for a comparison function u and all admissible h, then, putting $h = v - u$, with v an arbitrary comparison function, we obtain

(5) $$D[\hat{v}; \mathfrak{R}] = D[\hat{u}; \mathfrak{R}] + D[h; \mathfrak{R}] \geq D[\hat{u}; \mathfrak{R}],$$

so that $D[\hat{u}; \mathfrak{R}]$ must be the minimal value μ. This completes the proof.

Theorem 4 implies that for a given covering $\mathfrak{R}_0$, $\mathfrak{R}_1$, ... the comparison function u is uniquely determined if $D[\hat{u}; \mathfrak{R}] = \mu$ and $u - q$ vanishes at $\mathfrak{z}_0$. In fact, if v has these properties then the difference $v - u = h$ vanishes at $\mathfrak{z}_0$ and, since u has a minimum integral, we have (4) and therefore (5). But then we have, in succession,

$$D[h; \mathfrak{R}] = 0, \qquad h = 0, \qquad u = v.$$

Next we show that if the disks $\mathfrak{R}_1$, $\mathfrak{R}_2$, ... are held fixed, then u is independent of the choice of the disk $\mathfrak{R}_0$. We replace $\mathfrak{R}_0$ by a concentric disk $\mathfrak{R}^*$ of radius $a_0 > a$ which, like $\mathfrak{R}_0$, lies in the interior of $\mathfrak{R}_1$. We put, correspondingly,

$$q^*(\mathfrak{z}) = q^*(x, y) = \frac{x}{x^2 + y^2} + \frac{x}{a_0^2},$$

and $u^* = u - q^*$ in $\mathfrak{R}^*$, $u^* = u$ in $\mathfrak{R} - \mathfrak{R}^*$. In view of Theorem 4 we need only show that for all admissible h, (4) implies the corresponding relation

$$D[u^*, h; \mathfrak{R}] = 0.$$

This means that we must deduce the formula

$$D[\hat{u} - u^*, h; \mathfrak{R}] = 0.$$

Now

$$u - u^* = \begin{cases} 0 \text{ in } \mathfrak{R} - \mathfrak{R}^* \\ q^* \text{ in } \mathfrak{R}^* - \mathfrak{R}_0, \\ q^* - q \text{ in } \mathfrak{R}_0 \end{cases}$$

so that

$$D[\hat{u} - u^*, h; \mathfrak{R}] = D[q^* - q, h; \mathfrak{R}^*] + D[q, h; \mathfrak{R}^* - \mathfrak{R}_0].$$

We apply Green's identity to both summands on the right. The difference $q^* - q$ is harmonic in $\mathfrak{R}^*$, and the function q is harmonic in $\mathfrak{R}^* - \mathfrak{R}_0$. In step II of the proof of Theorem 3 we showed, in consequence of (3), that the expression $q_x \, dy - q_y \, dx$ vanishes on the boundary C_0 of $\mathfrak{R}_0$; but then $q_x^* \, dy - q_y^* \, dx$ vanishes on the boundary C^* of the disk $\mathfrak{R}^*$. It follows that

$$D[q^* - q, h; \mathfrak{R}^*] = -\int_{C^*} h(q_x \, dy - q_y \, dx),$$

$$D[q, h; \mathfrak{R}^* - \mathfrak{R}_0] = \int_{C^*} h(q_x \, dy - q_y \, dx),$$

and this yields our assertion. We note that, as is easily calculated, the value μ of the minimum depends on the radius of the disk $\mathfrak{R}_0$.

Finally we show that the extremum function u is independent of the choice of the remaining covering disks $\mathfrak{R}_1, \mathfrak{R}_2, \dots$. In fact, if v is the extremum function associated with another covering with the same center as $\mathfrak{R}_1$ then, by what has just been proved, we can use the same disk $\mathfrak{R}_0$ for both coverings. We put $h = v - u$ and use Theorem 4 with u, h and with v, $-h$ in place of u, h. It follows that

$$D[\hat{v}; \mathfrak{R}] = D[\hat{u}; \mathfrak{R}] + D[h; \mathfrak{R}], \qquad D[\hat{u}; \mathfrak{R}] = D[\hat{v}; \mathfrak{R}] + D[h; \mathfrak{R}],$$

and so

$$D[h; \mathfrak{R}] = 0, \qquad h = 0, \qquad v = u.$$

The harmonic conjugate $v(\mathfrak{z})$ of the extremum function $u(\mathfrak{z})$ is determined, as in the proof of Theorem 2, Section 5, by the line integral

$$v(\mathfrak{z}) - c = \int_L (u_x \, dy - u_y \, dx)$$

with c an arbitrary real constant. Here L is a rectifiable curve on $\overset{\cdot}{\mathfrak{R}}$ from a fixed point $\mathfrak{z}_1$ to a variable point $\mathfrak{z}$, and integration is with respect to the appropriate local uniformizing parameters. Now, in general, $\overset{\cdot}{\mathfrak{R}}$ is not simply connected; not even if $\mathfrak{R}$ is. We wish to prove that if $\mathfrak{R}$ is simply connected then $v(\mathfrak{z})$ is single valued. Let $\mathfrak{z}_1$ be a point of $\overset{\cdot}{\mathfrak{R}}_0$ and let L be closed. Then L is homotopic to a curve in $\mathfrak{R}_0$ which loops $\mathfrak{z}_0$, and all that needs to be proved is that the integral over such a curve has the value 0. We have $u = (u - q) + q$. The summand $u - q$ is harmonic on $\mathfrak{R}_0$ and has therefore a single-valued harmonic conjugate there. The summand q, being the real part of the analytic function $z^{-1} + za^{-2}$, has in $\overset{\cdot}{\mathfrak{R}}_0$ the single valued harmonic conjugate $-y(x^2 + y^2)^{-1} + ya^{-2}$. It follows that we may put

$$v(\mathfrak{z}) = v(x, y) = \frac{-y}{x^2 + y^2} + \frac{y}{a^2} + \int_0^z [(u - q)_x \, dy - (u - q)_y \, dx],$$

which incidentally determines at the same time the constant c. If we now put

$$f(\mathfrak{z}) = u(\mathfrak{z}) + iv(\mathfrak{z}),$$

then we obtain a function which is regular on $\overset{\cdot}{\mathfrak{R}}$ and actually single-valued there provided that the Riemann region is simply connected. The point $\mathfrak{z}_0$ is a simple pole with residue 1, and the difference $f(\mathfrak{z}) - z^{-1}$ vanishes at $\mathfrak{z} = \mathfrak{z}_0$.

Using the extremum function obtained above we can easily construct an analytic function whose Riemann surface is precisely $\mathfrak{R}$. Let $g(\mathfrak{z})$ be any single-valued meromorphic function on $\mathfrak{R}$ and let $\mathfrak{z}^*$ be a fixed point of $\mathfrak{R}$ with projection z^*. We consider an arbitrary path L on the z-sphere issuing from z^* and assume that the function element of $g(\mathfrak{z})$ defined in the vicinity of $\mathfrak{z}^*$ can by continued analytically along this path; here we admit poles and use the local parameters. If $\mathfrak{R}$ is precisely the Riemann surface of $g(\mathfrak{z})$, then such a path L must always be the projection of a path W on $\mathfrak{R}$ issuing from $\mathfrak{z}^*$. Further, if L is a closed path such that analytic continuation of the original function element along L leads back to that element, then W must be closed. Conversely, if both of these conditions are fulfilled then $\mathfrak{R}$ is the Riemann surface of $g(\mathfrak{z})$. It remains to construct a function $g(\mathfrak{z})$ of this kind.

We consider the single-valued function on $\mathfrak{R}$ given by

$$g_0(\mathfrak{z}) = \frac{df(\mathfrak{z})}{dz} = u_x(\mathfrak{z}) - iu_y(\mathfrak{z}).$$

This function has a double pole at $\mathfrak{z}_0$ with principal part $-z^{-2}$ and, except for the branch points, it is regular at all other finite points of $\mathfrak{R}$. If t is the local uniformizing parameter at a finite point or at a point at infinity, then the derivative (dz/dt) is regular there or it has a pole with respect to the parameter in question and, since the derivative $[df(\mathfrak{z})/dt]$ is regular there, it follows that $g_0(\mathfrak{z})$ is meromorphic throughout $\mathfrak{R}$. Since $\mathfrak{z}_0$ was an arbitrary point on $\mathfrak{R}$ whose projection was moved to 0 by means of a translation, and since there are finitely many or countably many branch points on $\mathfrak{R}$, it is possible to choose $\mathfrak{z}_0$ so that no branch point has the same projection $z = 0$ as $\mathfrak{z}_0$. Now let $\mathfrak{R}$ be compact and let L be a path on the z-sphere issuing from $z = 0$ along which it is possible to continue analytically the function element of $g_0(\mathfrak{z})$ given at $\mathfrak{z}_0$. If, beginning at $\mathfrak{z}_0$, we follow on $\mathfrak{R}$ the corresponding path over L then, in view of the compactness of $\mathfrak{R}$, we obtain for all of L a path W on $\mathfrak{R}$ lying over L. This shows that the first condition of the preceding paragraph is satisfied. Further, let L be a closed path on the z-sphere and let $\tilde{\mathfrak{z}}_0$ be the terminal point of W. Since $\tilde{\mathfrak{z}}_0$ lies over $z = 0$, it follows that this point is a double pole of the function $g_0(\mathfrak{z})$ viewed as a function of z only if $\tilde{\mathfrak{z}}_0 = \mathfrak{z}_0$, so that W is closed. We see that the second condition is also satisfied, and thus the function $g(\mathfrak{z}) = g_0(\mathfrak{z})$ satisfies all the requirements.

If $\mathfrak{R}$ is not compact then the preceding simple construction does not immediately yield the desired result. In this case we make use of the covering in Theorem 2 to determine on $\mathfrak{R}$ countably many points $\mathfrak{y}_1, \mathfrak{y}_2, \ldots$ as follows: For $t = 1, 2, \ldots$ we choose on the part of $\mathfrak{R}_{t+1}$ not covered by $\mathfrak{B}_t$ a sequence of finitely many interior points $\mathfrak{q}_t$ such that, with distances measured in the local parameters, every point of that part is $< t^{-1}$ away from at least one of the points of $\mathfrak{q}_t$. The traces of the points $\mathfrak{q}_1, \mathfrak{q}_2, \ldots$ are to be distinct, different from the projections of the branch points, and finite. The sequence $\mathfrak{y}_1, \mathfrak{y}_2, \ldots$ is the union of the finite sequences $\mathfrak{q}_1, \mathfrak{q}_2, \ldots$. Now let $u_n(\mathfrak{z})$ be the extremum function u with $\mathfrak{y}_n$ in place of $\mathfrak{z}_0$. This means that $u_n(\mathfrak{z})$ is harmonic on the Riemann region $\mathfrak{R}$ punctured at $\mathfrak{y}_n$ and has the appropriate singularity at $\mathfrak{y}_n$. If $\mathfrak{y}_n$ is one of the points of $\mathfrak{q}_t$, then $u_n(\mathfrak{z})$ is harmonic on $\mathfrak{B}_t$ and therefore bounded there. With m_n an upper bound for the absolute value of $u_n(\mathfrak{z})$ on $\mathfrak{B}_t$, the series

$$w(\mathfrak{z}) = \sum_{n=1}^{\infty} \frac{u_n(\mathfrak{z})}{n^2 m_n}$$

converges at all points $\mathfrak{z}$ of $\mathfrak{R}$ other than the points $\mathfrak{y}_1, \mathfrak{y}_2, \ldots$, and this convergence is uniform on every region $\mathfrak{B}_t$ $(t = 1, 2, \ldots)$. By Harnack's theorem, $w(\mathfrak{z})$ is harmonic on $\mathfrak{R}$ except at the points $\mathfrak{y}_1, \mathfrak{y}_2, \ldots$. The function

$$g(\mathfrak{z}) = w_x - i w_y$$

is meromorphic on $\mathfrak{R}$, has double poles at all the points $\mathfrak{y}_1, \mathfrak{y}_2, \ldots$, and no additional pole whose projection coincides with the projection of a point $\mathfrak{y}_k$ $(k = 1, 2, \ldots)$. Now let L be a path on the z-sphere issuing from the trace 0 of $\mathfrak{z}_0$ along which it is possible to continue analytically the given function

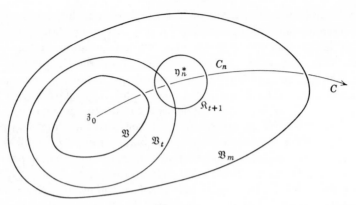

Figure 65

element of $g(\mathfrak{z})$, and suppose there is no path W on $\mathfrak{R}$ issuing from $\mathfrak{z}_0$ over all of the projected curve L. Then there is a maximal subarc B of L with starting point 0 and without terminal point over which there lies a corresponding arc C on $\mathfrak{R}$. The arc C cannot lie on a compact part of $\mathfrak{R}$, for otherwise B could be made longer. In the exterior of every region $\mathfrak{B}_n$ it is possible to find a closed subarc C_n of C which is covered by a region $\mathfrak{B}_m$. If we choose m to be as small as possible and put $t = m - 1$, then $t \geq n$ and the part of $\mathfrak{R}_{t+1}$ not covered by $\mathfrak{B}_t$ intersects C_n (Figure 65). It follows that there is a pole $\mathfrak{y}_n^*$ of $g(\mathfrak{z})$ on $\mathfrak{R}_{t+1}$ whose distance from C_n is less than t^{-1}. But then there would be projections of infinitely many poles arbitrarily close to L, and this contradicts the assumed possibility of continuation along the whole curve L. This proves that the function $g(\mathfrak{z})$ satisfies the first condition; that it satisfies the second condition is proved in much the same way as in the case of a compact $\mathfrak{R}$.

By Theorem 1 in Section 2, every compact Riemann region $\mathfrak{R}$ is of algebraic type. This means that $\mathfrak{R}$ can be decomposed by means of a finite number of cuts into a certain number n of sheets over the z-sphere. In this way the function $g_0(\mathfrak{z})$ is split into n branches $g_k(z)$ $(k = 1, \ldots, n)$ whose elementary symmetric polynomials are meromorphic on the z-sphere and so are rational functions of z. It follows that $g(\mathfrak{z})$ is an algebraic function of z whose Riemann surface is just $\mathfrak{R}$. This shows that every compact Riemann region is actually the Riemann surface of an algebraic function.

9. The mapping theorem

In this section we assume that $\mathfrak{R}$ is a simply connected Riemann region and study the mapping from $\mathfrak{R}$ into the w-sphere effected by the meromorphic function $w = f(\mathfrak{z})$.

Theorem 1: The mapping $w = f(\mathfrak{z})$ is conformal at all points of $\mathfrak{R}$ relative to the local parameters.

Proof: The function w^{-1} has a simple zero at $\mathfrak{z} = \mathfrak{z}_0$. This means that f maps a neighborhood of $\mathfrak{z}_0$ conformally onto a neighborhood of $w = \infty$. It remains to show that in all points of $\mathfrak{R}$ the derivative of $f(\mathfrak{z})$ with respect to the local parameter is different from zero. The proof is in two steps.

I.   In the neighborhood $\mathfrak{R}_1$ of $\mathfrak{z}_0$ we have the expansions

$$u + iv = w = \frac{1}{z} + c_1 z + \cdots ,$$

$$u = \frac{x}{x^2 + y^2} + a_1 x + \cdots , \qquad v = \frac{-y}{x^2 + y^2} + b_1 y + \cdots .$$

Let c be any real constant. The equation $v = c$ determines in the w-plane a parallel to the u-axis which we think of as running from right to left, that is from $u = \infty$ to $u = -\infty$. For all sufficiently large values of $|c|$ these parallels are the images under the given mapping (which we know to be locally conformal at $\mathfrak{z} = \mathfrak{z}_0$) of small simple closed curves touching the x-axis at $z = 0$ and approximating the circles given by the equation

$$c(x^2 + y^2) = -y.$$

The problem is to trace the curves on $\mathfrak{R}$ which are mapped by $w = f(z)$ onto the various lines $v = c$ for arbitrary c. If, for a fixed c, u^{-1} varies over a small interval about 0 then

$$z = w^{-1} + c_1 w^{-3} + \cdots = (u + ic)^{-1} + \cdots = u^{-1} - icu^{-2} + \cdots$$

varies over a small arc which touches the x-axis at the origin (since the sign of $Im\ z$ is constant). It follows that this arc divides a sufficiently small disk on $\mathfrak{R}$ with center $\mathfrak{z}_0$ into just two domains. Let $\mathfrak{H}$ denote the lower of these domains. Then $v > c$ in the lower domain $\mathfrak{H}$ and $v < c$ in the upper domain (Figure 66) since, as c increases, $-cu^{-2}$ decreases. We now consider all the points of $\mathfrak{R}$ at which $v(\mathfrak{y}) > c$ and claim that these form a single domain. To

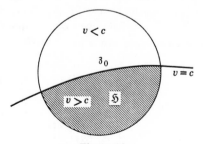

Figure 66

justify this claim we must prove that every such point can be joined to a point $\mathfrak{y}_1$ of $\mathfrak{H}$ by means of a curve on $\mathfrak{R}$ at all of whose points we have $v > c$. If this is false then there is a point $\mathfrak{y}_2$ on $\mathfrak{R}$ with $v(\mathfrak{y}_2) > c$ which fails to have the property in question. Let $\mathfrak{G}_k$ ($k = 1, 2$) be the set of all points of $\mathfrak{R}$ which can be joined to $\mathfrak{y}_k$ by means of curves for which $v > c$. These two sets form two disjoint domains with $\mathfrak{z}_0$ a boundary point of $\mathfrak{G}_1$ but not of $\mathfrak{G}_2$. We now produce a contradiction with the aid of Theorem 4 in Section 8.

Let $g(v)$ be a real-valued function which is continuously differentiable for all finite $v \geq 0$, bounded together with its derivative, positive for positive v, and which satisfies the additional conditions $g(0) = 0$, $g'(0) = 0$. One such function is

$$g(v) = \frac{v^2}{1 + v^2} = 1 - \frac{1}{1 + v^2}.$$

Further, let $k(u)$ be a real-valued function which is bounded and continuously differentiable for all real u, and whose derivative is bounded and positive for all u. One such function is

$$k(u) = \text{arc tan } u,$$

with

$$-\frac{\pi}{2} < \text{arc tan } u < \frac{\pi}{2}.$$

If we put

$$h(\mathfrak{z}) = k(u)g(v - c) = kg$$

for all $\mathfrak{z}$ in $\mathfrak{G}_2$ and $h(\mathfrak{z}) = 0$ in $\mathfrak{R} - \mathfrak{G}_2$, then the function $h(\mathfrak{z})$ is continuously differentiable in the local parameters throughout $\mathfrak{R}$. In $\mathfrak{G}_2$ we have

$$h_x = k'gu_x + kg'v_x = k'gu_x - kg'u_y,$$
$$h_y = k'gu_y + kg'v_y = k'gu_y + kg'u_x,$$
$$u_xh_x + u_yh_y = k'g(u_x^2 + u_y^2),$$
$$h_x^2 + h_y^2 = [(k'g)^2 + (kg')^2](u_x^2 + u_y^2),$$

and in $\mathfrak{R} - \mathfrak{G}_2$ both functions h_x and h_y vanish. The boundedness of the factors $(k'g)^2 + (kg')^2$ and the convergence of the normalized Dirichlet integral $D[\hat{u}; \mathfrak{R}]$ imply the existence of $D[h; \mathfrak{R}]$. Since the extremum function u is independent of the choice of $\mathfrak{R}_0$, we may assume that $\mathfrak{R}_0$ lies entirely in $\mathfrak{R} - \mathfrak{G}_2$; this is so because $\mathfrak{z}_0$ is not a point of $\mathfrak{G}_2$ or of its boundary. We then have

$$D[\hat{u}, h; \mathfrak{R}] = D[u, h; \mathfrak{G}_2] = \iint_{\mathfrak{G}_2} k'(u)g(v - c)(u_x^2 + u_y^2)\,dx\,dy,$$

where

$$u_x^2 + u_y^2 = \left|\frac{df}{dz}\right|^2.$$

It follows that the integrand is positive with the possible exception of isolated points of $\mathfrak{G}_2$. This implies the inequality

$$D[\hat{u}, h; \mathfrak{R}] > 0,$$

which contradicts Theorem 4 in the previous section. This proves that the inequality $v > c$ defines a single domain on $\mathfrak{R}$. If in the above argument we replace $g(v - c)$ with $g(c - v)$, then we can prove that the inequality $v < c$ also defines a single domain on $\mathfrak{R}$.

II.   Now we can show that on $\mathfrak{R}$ the derivative of $f(\mathfrak{z})$ with respect to the local parameter is everywhere $\neq 0$. We again use an indirect argument and consider a point $\mathfrak{z} = \mathfrak{z}^*$ at which $f(\mathfrak{z}) = f(\mathfrak{z}^*)$, viewed as a function of the local parameter t, vanishes to order $l > 1$. It follows that $w = f(\mathfrak{z})$ maps a schlicht neighborhood of $t = 0$ onto a neighborhood of $w^* = f(\mathfrak{z}^*)$ covered l times. This means that, as we pass the point $w^* = u^* + iv^*$ in the w-plane on the parallel $v = v^*$ to the u-axis, the inverse mapping from w to t yields as image of this parallel (on $\mathfrak{R}$ in the vicinity of $\mathfrak{z}^*$) precisely l curves which intersect at $\mathfrak{z}^*$ and form there $2l$ sectors with angular separation π/l. With these sectors arranged in cyclic order we have in their interiors, alternately, $v > v^*$ and $v < v^*$. We choose two points $\mathfrak{y}_1, \mathfrak{y}_3$ in disjoint sectors with $v > v^*$ and two points $\mathfrak{y}_2, \mathfrak{y}_4$ in disjoint sectors with $v < v^*$. In view of part I of this proof there is a curve L_1 in $\mathfrak{R}$ joining $\mathfrak{y}_1$ to $\mathfrak{y}_3$ at all of whose points we have $v > v^*$, and a curve L_2 in $\mathfrak{R}$ joining $\mathfrak{y}_2$ to $\mathfrak{y}_4$ at all of whose points we have $v < v^*$. Finally we join $\mathfrak{y}_1$ and $\mathfrak{y}_3$ in their sectors to $\mathfrak{z}^*$ and, similarly, $\mathfrak{y}_2$ and $\mathfrak{y}_4$ to $\mathfrak{z}^*$. In this way there arise two closed curves L_1 and L_2 on $\mathfrak{R}$ which, for proper choice of the four sectors, cross at $\mathfrak{z}^*$ (Figure 67). The curves L_1 and L_2 have no other point in common for, disregarding $\mathfrak{z}^*$, we have $v > v^*$ on one curve and $v < v^*$ on the other. We now show that this disposition of the curves L_1 and L_2 contradicts the assumed simple-connectedness of $\mathfrak{R}$. To this end we introduce the concept of the index

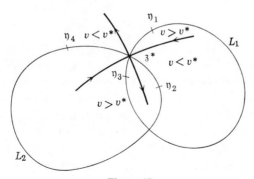

Figure 67

(A, B) of two curves A and B. This concept is also used in Chapter 4 (Vol. II).

Let A and B be two oriented curves on $\Re$ which cross in finitely many points $\mathfrak{q}_1, \ldots, \mathfrak{q}_m$. If $\mathfrak{q}$ is a point at which B crosses A in the same direction in which the y-axis crosses the x-axis, then we assign to $\mathfrak{q}$ the value 1. In the opposite case we assign to $\mathfrak{q}$ the value -1. By the *index* (A, B) we mean the sum of the values ± 1 extended over all the $\mathfrak{q} = \mathfrak{q}_1, \ldots, \mathfrak{q}_m$. Now let A and B be closed. Then it can be shown that the index is invariant under continuous deformation of A and B provided that the curves change in such a way that the number of points they have in common remains finite. Since $\Re$ is presumed simply-connected, every closed curve on $\Re$ is homotopic to zero; in particular, this is true of L_2. It is easy to see that L_2 can be contracted on on $\Re$ to a point outside L_1 in such a way that during the contraction process the number of points of intersection with L_1 stays finite. In the beginning of the deformation process we have $(L_1, L_2) = \pm 1$ and at its conclusion we have $(L_1, L_2) = 0$. This contradiction proves that, as asserted, the derivative of $f(\mathfrak{z})$ with respect to the local parameter does not vanish on $\Re$.

In the next section, Theorem 1 is used only in the case when $\Re$ is the covering surface of a compact Riemann surface. In that case, bearing in mind the fact (established in Section 3) that we can map such a $\Re$ topologically onto a simply connected schlicht domain in the plane or on the sphere, we can prove that the closed curves L_1 and L_2 cannot cross in just one point, using the Jordan curve theorem, and without introducing the notion of index. We can assume that the curves L_1 and L_2 are simple, closed, and piecewise smooth. The image of L_1 in the plane or on the sphere divides the plane into two disjoint parts. In traversing the image of L_2 we intercept both parts without intercepting the image of L_1. This then yields a new proof of the result of the previous section for the case when $\Re$ is the covering surface of a compact Riemann surface.

Let L be a curve on $\Re$ issuing from $\mathfrak{z}_0$ which is mapped by $w = f(\mathfrak{z})$ onto a curve C on the w-sphere issuing from $w = \infty$. Further, let

$$z = \zeta(w) = w^{-1} + c_1 w^{-3} + \cdots$$

be the function element at $w = \infty$ obtained by inverting $w = f(\mathfrak{z})$ at the point $\mathfrak{z} = \mathfrak{z}_0$. In view of Theorem 1, $\zeta(w)$ can be continued analytically along C, and this analytic continuation maps a sufficiently small neighborhood of every point on C conformally onto a domain in $\Re$. In the converse case of a prescribed C we wish to continue the function element $\zeta(w)$ analytically along the line $v = c$ where u can be taken to increase from $-\infty$ or to decrease from ∞. For a given c we distinguish two cases:

In the first case analytic continuation is possible on the whole line $v = c$ from $u = -\infty$ to $u = \infty$, and the line is mapped onto some curve L in

$\Re$. Since $f(\mathfrak{z})$ is single-valued on $\Re$ and infinite only at $\mathfrak{z} = \mathfrak{z}_0$, it follows that L must be a simple closed curve issuing from $\mathfrak{z}_0$, and that after traversing the whole line we arrive at $u = \infty$ with the initial function element $\zeta(w)$. The same is true if we traverse $v = c$ in the opposite direction.

In the second case there is a largest real value a such that the ray $u < a$, $v = c$ is the image under $w = f(\mathfrak{z})$ of a curve L_1 on $\Re$ issuing from $\mathfrak{z}_0$. Then, correspondingly, there is also a smallest real value b such that the ray $u > b$, $v = c$ is the image of a curve L_2 on $\Re$ issuing from $\mathfrak{z}_0$. Surely, the function element $\zeta(w)$ can be continued analytically along these two open rays beginning at $w = \infty$. It is conceivable that the analytic continuation can be carried out from the left up to the point $u = a, v = c$ or from the right up to the point $u = b, v = c$, but then the image under $z = \zeta(w)$ of the corresponding closed ray on the w-sphere is no longer the projection of a curve on $\Re$ issuing from $\mathfrak{z}_0$.†

In view of the single-valuedness of $f(\mathfrak{z})$ on $\Re$, neither L_1 nor L_2 has a double point. If L_1 and L_2 had a common point $\mathfrak{z}^* \neq \mathfrak{z}_0$, then the real part d of $f(\mathfrak{z}^*)$ would satisfy the inequality $b < d < a$. In that case, the analytic continuation of $\zeta(w)$ along $v = c$ up to $u = d$ would lead on both paths to the same function element, namely, to the power series which is the inverse of $f(\mathfrak{z})$ at $\mathfrak{z}^*$. But this would result in a contradiction, for it would imply the possibility of analytic continuation beyond a for increasing u. Hence L_1 and L_2 form together a simple curve L on $\Re$ containing $\mathfrak{z}_0$. We now want to see in greater detail how the image $\mathfrak{z}$ of a point of the ray $u < a, v = c$ moves on L_1 when u approaches a. Let $\mathfrak{M}$ be any compact subset of $\Re$. We claim that it is possible to associate with $\mathfrak{M}$ a positive number δ such that for $a - \delta < u < a$ the point $\mathfrak{z}$ lies outside $\mathfrak{M}$. If this were not the case then there would exist a monotonically increasing sequence u_n ($n = 1, 2, \ldots$) converging to a such that the sequence of images $\mathfrak{y}_n$ on $\mathfrak{M}$ converges to a point $\mathfrak{q} \neq \mathfrak{z}_0$. In view of the continuity of $f(\mathfrak{z})$ on $\Re$ we would have $f(\mathfrak{q}) = a + ci$, and it would be possible to continue $\zeta(w)$ analytically along $v = c$ from the left up to the point $u = a$, namely, as the inverse function of $f(\mathfrak{z})$ in the vicinity of $\mathfrak{z} = \mathfrak{q}$, a possibility ruled out by the assumed maximum property of a. This proves the assertion concerning the existence of δ. In turn, this assertion implies that if u approaches a from the left then the image $\mathfrak{z}$ on L_1 leaves eventually each of the compact regions $\mathfrak{B}_n$ ($n = 1, 2, \ldots$). The same is true of L_2 when u decreases and converges to b. With these properties of the subarcs L_1 and L_2 in mind we shall refer to L as an *endless curve*.

Theorem 2:   An endless curve occurs for at most one value of c.

Proof: Let $c < c^*$ be two real numbers and let L and L^* be the endless

† $\Re$ may not be the complete Riemann surface defined by the function element of f at z_0. (Tr.)

curves with subarcs L_1, L_2 and L_1^*, L_2^* associated with these numbers. In view of the single-valuedness of $f(\mathfrak{z})$, the curves L and L^* have only the point $\mathfrak{z}_0$ in common, at which point they touch the x-axis without crossing one another. If $\mathfrak{K}$ is a sufficiently small disk about $\mathfrak{z}_0$, then L_1, L_1^* and L_2, L_2^* determine on $\mathfrak{K}$ two sectors $\mathfrak{S}_1$ and $\mathfrak{S}_2$ with zero angular separation. We claim that every interior point $\mathfrak{y}_1$ of $\mathfrak{S}_1$ can be joined to every interior point $\mathfrak{y}_2$ of $\mathfrak{S}_2$ by means of a curve in $\mathfrak{K}$ for all of whose points we have $c < v < c^*$. This is proved indirectly as follows:

With $-m$ a sufficiently large positive number we consider the parallel to the v-axis defined by the equation $u = m$ and its image C on $\mathfrak{K}$ under the power series $z = \zeta(w)$. The curve C is then approximately a small circle inside $\mathfrak{K}$.

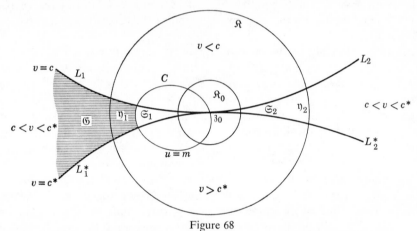

Figure 68

C is orthogonal to the x-axis at $\mathfrak{z}_0$ and it cuts off an even smaller sector of $\mathfrak{S}_1$ without otherwise intersecting $\mathfrak{S}_2$. Let $\mathfrak{y}_1$ be a point in the interior of $\mathfrak{S}_1$ lying outside C. We consider on $\mathfrak{K}$ the totality of points with $u > m$ and $c < v < c^*$. These may form a number of disjoint domains of which we pick the domain $\mathfrak{G}$ containing $\mathfrak{y}_1$. If $\mathfrak{y}_2$ were not a point of $\mathfrak{G}$ then, in particular, $\mathfrak{z}_0$ would not be a boundary point of $\mathfrak{G}$ on $\mathfrak{K}$. We can therefore assume that $\mathfrak{K}_0$ lies in $\mathfrak{K} - \mathfrak{G}$ (Figure 68). If the function $g(v)$ satisfies the requirements stated in the proof of Theorem 1, with

$$g(v) = \frac{v^2}{1 + v^2}$$

an admissible choice, then, in the present argument, we define $h(\mathfrak{z}) = g(u - m)\, g(v - c)\, g(c^* - v)$ in $\mathfrak{G}$ and $h(\mathfrak{z}) = 0$ on $\mathfrak{K} - \mathfrak{G}$. Putting, for brevity,

$$g'(u - m)g(v - c)g(c^* - v) = g_1,$$
$$g(u - m)(g'(v - c)g(c^* - v) - g(v - c)g'(c^* - v)) = g_2,$$

we have in $\mathfrak{G}$ the equalities

$$h_x = g_1 u_x - g_2 u_y, \qquad h_y = g_1 u_y + g_2 u_x,$$
$$u_x h_x + u_y h_y = g_1(u_x^2 + u_y^2), \qquad h_x^2 + h_y^2 = (g_1^2 + g_2^2)(u_x^2 + u_y^2).$$

This again implies the existence of the Dirichlet integral $D[h; \mathfrak{R}]$, and we have $D[\hat{u}, h; \mathfrak{R}] > 0$. This leads to the contradiction encountered earlier in the proof of Theorem 1. It follows that $\mathfrak{y}_2$ lies in $\mathfrak{G}$, and we can join $\mathfrak{y}_1$ and $\mathfrak{y}_2$ by means of a curve in $\mathfrak{R}$ on which we have, invariably, $c < v < c^*$. Finally, by joining $\mathfrak{y}_1$ to $\mathfrak{z}_0$ by means of a curve on $\mathfrak{S}_1$ and $\mathfrak{y}_2$ to $\mathfrak{z}_0$ by means of a curve on $\mathfrak{S}_2$, we obtain a closed curve B on $\mathfrak{R}$ issuing from $\mathfrak{z}_0$ on which for $\mathfrak{z} \neq \mathfrak{z}_0$ we have, invariably, $c < v < c^*$. On the other hand, let A be the endless

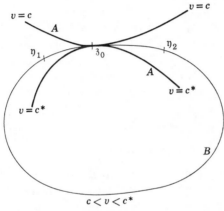

Figure 69

curve consisting of L_1 and L_2^* (Figure 69). Then A and B cross just in the one point $\mathfrak{z}_0$, and the index (A, B) is ± 1. In view of the simple-connectedness of $\mathfrak{R}$, B can be contracted to a point outside A in such a way that throughout the process of deformation the fixed endless curve A is invariably intercepted in only finitely many points. Since there are eventually no points of intersection and the index remains unchanged, we obtain a contradiction once again. In particular, if $\mathfrak{R}$ is the covering surface of a compact Riemann region, then the contradiction can be obtained by using the Jordan curve theorem. All in all, we cannot have two endless curves L and L^*, and this is the assertion of our theorem.

Now let L be an endless curve on which $v = c$, and let L_1 with $u < a$ and L_2 with $u > b$ be the two subarcs of L which meet at $\mathfrak{z}_0$. We shall show that $a \leq b$. Suppose we had $b < a$. We put $d = (a + b)/2$ and consider for any value $c_1 < c$ the strip $c_1 \leq v < c$. Since L is the only endless curve, it follows that the function element $z = \zeta(w)$ admits of unique analytic continuation to all the points of the strip (Figure 70). If we adjoin the rays L_1 and L_2 then

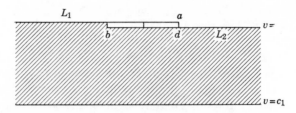

Figure 70

analytic continuation along these rays yields at d the same function elements. But then we would arrive at the familiar contradiction that continuation along L_1 beyond a is possible.

We see that there occur only the following three cases. There is no endless curve, there is an endless curve $v = c$ with $a = b$ or an endless curve with $a < b$. In the first case let $\mathfrak{W}$ be the full w- sphere, in the second case the sphere punctured at $w_1 = a + ci$, and in the third case the sphere slit along the segment $a \leq u \leq b$, $v = c$ from $w_1 = a + ci$ to $w_2 = b + ci$ with the edges of the slit omitted. We designate these three cases as elliptic, parabolic, and hyperbolic.

Theorem 3: The function $w = f(\mathfrak{z})$ effects a conformal and schlicht mapping of the simply connected Riemann region $\mathfrak{R}$ onto $\mathfrak{W}$.

Proof: Let $\mathfrak{y}_1$ be a point of $\mathfrak{R}$ other than $\mathfrak{z}_0$ and let c be the imaginary part of $f(\mathfrak{y}_1)$. We consider on $\mathfrak{R}$ the image curve L which corresponds to the line $v = c$ under the analytic continuation of the function element $z = \zeta(w)$. Then L is closed or endless, and on one side of L we have $v > c$ and on the other $v < c$. If $\mathfrak{y}_1$ were not on L, then we could find two points $\mathfrak{y}_2$ and $\mathfrak{y}_3$ sufficiently close to $\mathfrak{y}_1$ with $v(\mathfrak{y}_2) > c$, $v(\mathfrak{y}_3) < c$ which could be joined by means of a curve not meeting L. Since the inequalities $v(\mathfrak{z}) > c$, $v(\mathfrak{z}) < c$ define two domains on $\mathfrak{R}$, it would be possible to construct a closed curve B through the points $\mathfrak{z}_0$, $\mathfrak{y}_2$, $\mathfrak{y}_3$ crossing L at the single point $\mathfrak{z}_0$ (Figure 71).

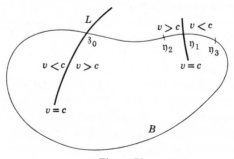

Figure 71

Consideration of (L, B), however, has shown this to be impossible. It follows that every point of $\mathfrak{R}$ lies on a curve L and, conversely, under the mapping $w = f(\mathfrak{z})$, for variable c, there correspond uniquely to the curves L all the parallel lines $v = c$, insofar as they belong to $\mathfrak{W}$. This means that the mapping covers $\mathfrak{W}$. Further, if $f(\mathfrak{z}) = f(\mathfrak{y}_1)$ then $\mathfrak{z}$ must also lie on L and, since every point on L is uniquely determined by the value of u, it follows that $\mathfrak{z} = \mathfrak{y}_1$. Hence the mapping is schlicht and the proof of our theorem is complete.

Theorem 3 shows that the inverse function $z = \zeta(w)$ is single-valued and meromorphic on $\mathfrak{W}$, that is, regular apart from poles. In the elliptic case $\mathfrak{W}$ is the full complex w-sphere, so that in this case the Riemann region $\mathfrak{R}$ is compact and, in fact, of genus 0. In the parabolic case we introduce the transformation $s = (w - w_1)^{-1}$ which maps the w-sphere punctured at $w = w_1$ conformally onto the full s-plane. In the hyperbolic case we put

$$ s = \frac{\sqrt{w - w_1} - \sqrt{w - w_2}}{\sqrt{w - w_1} + \sqrt{w - w_2}}, $$

where we have in mind the branch which satisfies the inequality $0 < s < 1$ for positive values of $w - w_2$ on the w-sphere slit from w_1 to w_2. In this way $\mathfrak{W}$ is mapped conformally onto the interior of the unit circle in the s-plane. Corresponding to the three cases we have mapped $\mathfrak{R}$ conformally onto the sphere, the plane and the interior of the unit circle. We denote these three possible domains by $\mathfrak{S}$. Of these only the sphere is compact, so that the elliptic case arises if and only if $\mathfrak{R}$ is compact. While the parabolic and hyperbolic cases are topologically indistinguishable, they are basically different in a function-theoretical sense for, by Liouville's theorem, it is not possible to map the complex plane conformally onto a bounded domain such as the unit disk. In this connection there arises the problem of deciding between the parabolic and hyperbolic cases by direct investigation of the given noncompact simply-connected Riemann region $\mathfrak{R}$. No generally applicable criterion of this type is known. In the sequel, in this and in the next sections we consider two important special cases.

Let $\mathfrak{R}$ be a schlicht and simply-connected domain on the z-sphere. If the boundary of $\mathfrak{R}$ is empty or consists of a single point then, trivially, we have the elliptic or parabolic case. Now let z_1 and z_2 be two distinct boundary points of $\mathfrak{R}$ and let z_3 be an interior point. Let $\mathfrak{R}^*$ be the image of $\mathfrak{R}$ under the fractional linear transformation which carries the points z_1, z_2, z_3 into the points $\infty, 0, 1$, and consider on $\mathfrak{R}^*$ the function $t = \sqrt{z}$, with $t = 1$ for $z = 1$. Since $\mathfrak{R}^*$ is simply connected and does not contain the branch points $\infty, 0$ of t, it follows by the monodromy theorem that t is single-valued there and effects a conformal mapping of $\mathfrak{R}^*$ onto a schlicht and simply-connected region $\mathfrak{R}^{**}$ in the t-plane. In view of the fact that we are dealing with the branch of $t = \sqrt{z}$ which maps a vicinity of $z = 1$ onto a vicinity of $t = 1$, a sufficiently small neighborhood of $t = -1$ lies outside $\mathfrak{R}^{**}$. If it were possible to map the full s-plane conformally onto $\mathfrak{R}^{**}$ then consideration of the function $(t + 1)^{-1}$ would again lead to a conclusion contradicting Liouville's theorem. We are therefore dealing with the hyperbolic case. This proves the Riemann mapping theorem stated in the beginning of Section 7.

10. Uniformization of algebraic functions

In this section we apply the results established thus far to the case of the covering surface of a compact Riemann region of genus p. As in Section 3 we denote the covering surface by $\mathfrak{U}$ and use $\mathfrak{R}$ to denote the corresponding compact Riemann region. With every element α of the fundamental group Γ of $\mathfrak{R}$ we can associate a conformal mapping α^* of $\mathfrak{U}$ onto itself as follows: Let A be a closed path on $\mathfrak{R}$ issuing from $\mathfrak{z}_0$ and belonging to the homotopy class α. If $\mathfrak{z}$ is any point on $\mathfrak{U}$ then we join $\mathfrak{z}_0$ to $\mathfrak{z}$ on $\mathfrak{U}$ by means of a path whose projection on $\mathfrak{R}$ is L. Further, let $\mathfrak{z}_\alpha$ on $\mathfrak{U}$ be the terminal point of the path with initial point $\mathfrak{z}$ lying over $L^{-1}AL$. The point $\mathfrak{z}_\alpha$ depends only on $\mathfrak{z}$ and α. Clearly, the correspondence α^* of $\mathfrak{U}$ onto itself which associates $\mathfrak{z}$ to $\mathfrak{z}_\alpha$ is conformal, for it sends every disk on $\mathfrak{U}$ into a congruent disk while the projected disk on $\mathfrak{R}$ remains unchanged. The relation

$$(L^{-1}AL)(L^{-1}BL) \sim L^{-1}(AB)L$$

implies

$$(\alpha\beta)^* = \alpha^*\beta^*, \qquad (\alpha^{-1})^* = (\alpha^*)^{-1};$$

moreover, ε^* is the identity, and $\alpha^* = \beta^*$ only for $\alpha = \beta$. This shows that the correspondence $\alpha \to \alpha^*$ defines a faithful representation of the fundamental group. The conformal maps α^* are called *covering transformations* of $\mathfrak{U}$ because the individual copies $\mathfrak{F}_\omega$ of the cut Riemann surface $\mathfrak{R}$ are carried by α^* into $\mathfrak{F}_{\alpha\omega}$ and so are permuted among themselves whereas the covering of $\mathfrak{U}$ by the $\mathfrak{F}_\omega$ remains unchanged as a whole. Since $L^{-1}AL$ is null homotopic only when $\alpha = \varepsilon$, we have $\mathfrak{z}_\alpha \neq \mathfrak{z}$ when $\alpha \neq \varepsilon$. Thus no point is left fixed by α^* when $\alpha \neq \varepsilon$.

We now replace w in the function $z = \zeta(w)$ obtained by inverting $w = f(\mathfrak{z})$ at $\mathfrak{z}_0$ by the new variable s defined at the end of the previous section, and obtain as a result $z = \psi(s)$. By Theorem 3 of that section, $z = \psi(s)$ effects a conformal mapping of the domain $\mathfrak{S}$ of the s-sphere onto the covering surface $\mathfrak{U}$ of $\mathfrak{R}$. Here $\mathfrak{S}$ is the full sphere in the elliptic case, the full s-plane in the parabolic case, and the interior of the unit circle in the hyperbolic case. If we denote the conformal mapping of $\mathfrak{U}$ onto $\mathfrak{S}$ by κ, then there corresponds to the covering transformation α^* the conformal mapping $\kappa^{-1}\alpha^*\kappa$ of the schlicht domain $\mathfrak{S}$ onto itself. Also, for $\alpha \neq \varepsilon$ this mapping carries no point of $\mathfrak{S}$ into itself and so has no fixed points on $\mathfrak{S}$.

Theorem 1: The covering surface of a compact Riemann surface of genus p is elliptic, parabolic or hyperbolic according to $p = 0$, $p = 1$, or $p > 1$.

Proof: The proof is accomplished in three steps.

I.   The fundamental group Γ of a compact Riemann surface $\mathfrak{R}$ of genus p consists in the case $p = 0$ of the identity ε alone, and has infinitely many elements α for $p > 0$. For $p = 0$ the covering surface $\mathfrak{U} = \mathfrak{R}$ is compact and we have the elliptic case. For $p > 0$ there are over every disk $\mathfrak{K}$ of $\mathfrak{R}$ in-finitely many congruent disjoint disks $\mathfrak{K}_\alpha$ whose centers have no limit point on $\mathfrak{U}$, which means that $\mathfrak{U}$ is not compact and we do not have the elliptic case. We conclude that the elliptic case occurs if and only if $p = 0$.

II.   In the parabolic case the covering transformations of $\mathfrak{U}$ yield a group of conformal mappings of the s-plane onto itself. If $t = g(s)$ is such a mapping, then, in particular, t is regular for all finite complex s, and so is an entire function. The same is true of the inverse function. Since the function $g(s)$ is not constant, it follows from Liouville's theorem that it is unbounded. If $g(0) = q$, then the unit disk of the s-plane is mapped onto a certain schlicht neighborhood $\mathfrak{U}$ of the point q, and for $|s| > 1$ the value $g(s)$ lies outside $\mathfrak{U}$; so that, in particular, the function $[g(s) - q]^{-1}$ is regular and bounded for $|s| > 1$. As such it can be uniquely continued up to the point $s = \infty$, and it is also regular there. It follows that $g(s)$ has a pole at $s = \infty$; therefore it is a polynomial. Now, in view of the conformal nature of the mapping, $g'(s)$ is different from 0 for all finite s and we conclude, by the fundamental theorem of algebra, that $g'(s)$ is constant and $g(s)$ is linear,

$$g(s) = cs + a \qquad (c \neq 0)$$

with constant c and a. If we had $c \neq 1$ then the solution s_0 of the equation

$$s_0 = cs_0 + a$$

would yield a finite fixed point of the mapping. Hence for the covering trans-formation under consideration, $c = 1$ and $g(s) = s + a$ with $a = 0$ in the case of the identity. For arbitrary complex a the translations $g(s) = s + a$ form a commutative group. Since the fundamental group Γ is not com-mutative for $p > 1$ and the covering transformations of $\mathfrak{U}$ yield a faithful representation of Γ, it follows that the parabolic case cannot occur for $p > 1$. Hence for $p > 1$ we always have the hyperbolic case.

III.   It remains to show that the hyperbolic case cannot arise for $p = 1$. In the hyperbolic case the covering transformations of $\mathfrak{U}$ yield a group of conformal mappings $t = g(s)$ of the unit disk $|s| < 1$ onto itself. First we consider the case when $g(0) = 0$. We use the mode of argument of the so-called Schwarz lemma and consider the function

$$h(s) = \frac{g(s)}{s} \qquad (|s| < 1)$$

which is regular in the unit disk. On every smaller concentric circular boundary we have

$$|h(s)| = \rho^{-1}|g(s)| \le \rho^{-1} \qquad (|s| = \rho; \; 0 < \rho < 1),$$

and, by the maximum principle, this inequality holds throughout the disk $|z| \le \rho$. Letting ρ increase to 1 we obtain

(1) $$|h(s)| \le 1 \qquad (|s| < 1).$$

Another application of the maximum principle makes it clear that if $h(s)$ is not a constant of absolute value 1, then we can leave out the equality sign in (1). With $h(0) = g'(0) = \lambda$ we see that, invariably, $|\lambda| \le 1$ and $|\lambda| = 1$ precisely when $h(s)$ is a constant of absolute value 1. We apply this estimate to the inverse function of $g(s)$. Then λ is replaced by λ^{-1} and we have $|\lambda^{-1}| \le 1$. It follows that

$$|\lambda| = 1, \qquad h(s) = \lambda, \qquad g(s) = \lambda s,$$

and, since $|\lambda| = 1$, $g(s)$ actually maps the unit disk conformally onto itself and leaves the origin fixed.

Now we consider the more general case $g(0) = a$, $|a| < 1$. In this case, the fractional linear transformation

$$w = \frac{t - a}{\bar{a}t - 1}, \qquad t = \frac{w - a}{\bar{a}w - 1}$$

effects a conformal mapping of the unit disk $|t| < 1$ onto itself and carries the point $t = a$ into the origin; this follows from the relation

$$|at - 1|^2 - |t - a|^2 = (1 - |t|^2)(1 - |a|^2) > 0 \qquad (|t| < 1).$$

If we put $t = g(s)$ then, in view of the previous section, we must have $|\lambda| = 1$ in the composite function $w = \lambda s$, and so

$$g(s) = t = \frac{\lambda s - a}{\bar{a}\lambda s - 1} \qquad (|\lambda| = 1, |a| < 1).$$

Replacing in this relation a by λa and then λ by $-\lambda$ we obtain

(2) $$g(s) = \lambda \frac{s - a}{1 - \bar{a}s} \qquad (|\lambda| = 1, |a| < 1).$$

This shows that all orientation preserving conformal mappings of the unit disk onto itself are given by the fractional linear functions of the form (2). These functions carry the boundary of the unit disk into itself and the exterior of the unit disk into itself.

In the case $p = 1$ the fundamental group Γ is the free abelian group on two generators α and β. If this were the hyperbolic case then α and β would be represented by two commuting transformations of the form (2). To simplify study we map the unit disk $|s| < 1$ onto the complex upper half plane by means of a suitable fractional linear transformation. In view of (2), all the orientation preserving conformal transformations of the upper half plane onto itself are also given by fractional linear transformations

$$(3) \qquad w^* = \frac{aw + b}{cw + d} \qquad (ad - bc \neq 0),$$

and since the real axis and its orientation are to be preserved, we may suppose a, b, c, d real and, consequently, $ad - bc > 0$. If such a transformation is not the identity, that is, if the equalities $a = d$, $b = c = 0$ do not occur at the same time, then it has on the w-sphere two simple conjugate complex fixed points, or two simple real fixed points, or one double real fixed point. In the present case the first possibility is ruled out since the mappings (3) representing α and β have no fixed points in the upper half plane. We may assume without restriction of generality that the mapping associated with α has the fixed points 0 and ∞ or the double fixed point ∞, and so is given by one of the normal forms

$$(4) \qquad w^* = kw \quad (k > 0, k \neq 1); \qquad w^* = w + l \quad (l^2 > 0),$$

while β is represented by (3). Corresponding to the two cases in (4), the relation $\alpha\beta = \beta\alpha$ implies

$$\frac{akw + b}{ckw + d} = k\frac{aw + b}{cw + d}, \qquad \frac{a(w + l) + b}{c(w + l) + d} = \frac{aw + b}{cw + d} + l$$

identically in w, so that

$$ack(k - 1) = bc(k^2 - 1) = bd(k - 1) = 0$$
$$ac = bc = bd = 0$$
$$b = c = 0$$
$$c^2l = (cl + 2d)cl = (cdl + d^2 + bc - ad)l = 0$$
$$c = d(d - a) = 0$$
$$c = a - d = 0.$$

It follows that the mapping associated with β has the normal form (4). All in all, α and β are represented either by the two transformations $w^* = k_1w$,

$w^* = k_2 w$ or by the two transformations $w^* = w + l_1$, $w^* = w + l_2$; here k_1 and k_2 are positive, l_1 and l_2 are real, $k_1 \neq 1$, $k_2 \neq 1$, $l_1 l_2 \neq 0$. Now we form for $\alpha^\xi \beta^\eta = \gamma$, ξ, η arbitrary integers, the associated linear transformations $w^* = k_1^\xi k_2^\eta w$ or $w^* = w + l_1 \xi + l_2 \eta$ and, for a given $\delta > 0$, choose the numbers ξ and η so that they both are not 0, and so that the appropriate one of the inequalities $|\xi \log k_1 + \eta \log k_2| < \delta$ or $|l_1 \xi + l_2 \eta| < \delta$ holds. Then $\gamma \neq \varepsilon$ and the appropriate linear transformation is different from the identity. If $\Re$ is a circle on $\Re$ with center $\mathfrak{z}$ then the disks $\Re_\varepsilon$ and $\Re_\gamma$ on $\mathfrak{U}$ are disjoint; whereas for sufficiently small δ the image of $\Re_\varepsilon$ in the upper half plane contains the image of $\mathfrak{z}_\gamma$. This contradiction proves that for $p = 1$ we have only the parabolic case. This completes the proof.

Now we consider an arbitrary function $q = q(z)$ which is single-valued and analytic on the compact Riemann region $\Re$. In place of the variable z in $q(z)$ we introduce the variable s defined by the conformal mapping $z = \psi(s)$ of $\mathfrak{S}$ onto the covering surface $\mathfrak{U}$. This makes $q = \chi(s)$ into an analytic function of s on $\mathfrak{S}$. We first put $p = 0$ so that $\mathfrak{S}$ is the full s-sphere and so $\chi(s)$ is a rational function of s. In particular, this holds for $q(z) = z$, that is, for the function $\psi(s)$ itself. Then the algebraic equation $z = \psi(s)$ defines s as an algebraic function of z which, in view of the equality $\mathfrak{U} = \Re$, is single-valued and, apart from one simple pole, regular throughout $\Re$. By an argument used before it follows that $\Re$ is the Riemann surface of s viewed as a function of z. It follows that in the case $p = 0$ the field of functions meromorphic on $\Re$ consists precisely of all the rational functions of s, and s is the inverse of the rational function which expresses z in terms of s.

Now let $p = 1$. Then there correspond to the generators α and β of the fundamental group the two translations $s^* = s + a$ and $s^* = s + b$ of the s-plane and, since $q(z)$ is supposed single-valued on $\Re$, $q = \chi(s)$ must be invariant under these translations. It follows that the meromorphic function $\chi(s)$ has the two independent periods a and b and is thus an elliptic function of s, and the ratio b/a of the periods is not real. Finally, if $p > 1$ then the fundamental group Γ is represented faithfully by the fractional linear substitutions

$$s^* = \frac{as + b}{cs + d}$$

which carry the unit disk $|s| < 1$ into itself, and

$$\chi(s^*) = \chi(s).$$

Hence the function $\chi(s)$, analytic in the unit disk, goes over into itself under all of the substitutions of this representation of Γ, and is therefore called an automorphic function. The automorphic functions are seen to be the natural analogs of the elliptic functions.

In all three cases the single-valued function $q(z)$ on $\mathfrak{R}$ is made into a single-valued function in the schlicht domain $\mathfrak{S}$ by introduction of the variable s. In the case of a compact Riemann region $\mathfrak{R}$ the variable s plays the same role in the large which the local uniformizing parameters play in the small. This explains the use of the term uniformization to describe the introduction of s. For a two-sheeted Riemann surface with four branch points the uniformizing parameter was obtained explicitly in the form of the elliptic integral of the first kind. In distinction to this the method used for the determination of the mapping function which occurs in Theorem 3 of Section 9 yields an existence proof and not a constructive algorithm. In the next chapter (Vol. II) we represent the automorphic functions explicitly by means of the Poincaré series whose construction is analogous to the construction of the partial fraction series for the elliptic functions $\wp(s)$ and $\wp'(s)$.

A few brief historical remarks. The basic ideas of the theory of uniformization are due to Klein and Poincaré who developed them independently in the eighties of the last century without giving rigorous proofs however. At about that time Schwarz conceived the fruitful idea of the covering surface of a Riemann surface. A complete proof of the uniformization theorem for algebraic functions and further generalizations were given in 1907 by Koebe and, at about the same time, by Poincaré.

Index

Abelian function, 104
Abelian integral, 103
Abelian integral of the first kind, 103
Algebraic function, 101
Analytic continuation of power series, 13
Analytic function, 22
Automorphic function, 104

Boundary value of function, 129
Branch number of point, 125
Branch number of Riemann region, 125
Branch point, 18, 21

Canonical dissection, 49
Comparison function, 158
Continuous function, 157
Covering surface, 40
Covering transformation, 178
Crosscut, 49
Curve, 16
Curve element, 16
Curve homotopic to zero, 38

Degenerate elliptic function, 85
Diagonalization, 110
Differentiable function, 157
Dirichlet integral, 139
Dirichlet principle, 145
Doubly periodic function, 42
Dual of polygon, 117

Elliptic function, 10
Elliptic integral, 11
Elliptic integral of the first kind, 11
Endless curve, 173
Euler characteristic, 125
Exponent sum of group element, 127

Function element, 16, 17
Fundamental group, 38

General elliptic integral, 79
Genus of compact Riemann region, 127
Green's identity, 141

Harmonic continuation, 132
Harmonic function, 130, 157
Homeomorphic objects, 29
Homeomorphism, 29
Homology group, 128
Homotopic curves, 37, 113

Index of pair of curves, 172
Intransitive group of permutations, 99

Legendre normal form, 57
Lemniscatic integral, 3
Linked surface elements, 16
Local parameter, 21
Local uniformizing parameter, 21

Meromorphic function, 101
Minimizing sequence, 158
Monodromy group of algebraic function,
   101

Normalized Dirichlet integral, 158

Order of an elliptic function, 72

Period of function, 40
Period parallelogram, 48
$\wp$-function, 59
Point lies above its projection, 16
Points at infinity, 18
Poisson kernel, 133
Poisson integral, 137
Poisson integral formula, 133
Puiseux series, 98
Punctured Riemann surface, 18

Ramification point, 18
Ramified surface, 20
Rearrangement of series, 12
Reduced pair of periods, 46
Regular analytic function, 22
Regular function, 17
Riemann domain, 16
Riemann region, 16

Riemann region of algebraic type, 105
Riemann surface, 18

Schwarz reflection principle, 152
Simply connected domain, region, 28
Smoothed minimizing sequence, 161
Sphere with p handles, 106
Sum of zeros of elliptic function, 72
Sum system of group element, 127
Surface element, 16

Topological mapping, 29
Transformation, 107
Transitive group of permutations, 99
T times smoothed minimizing sequence, 161

Unimodular substitution, 45

Weierstrass normal form, 58
Winding number, 19